W9-DBH-149

What Should We Do with Our Brain?

DISCARD

Series Board

James Bernauer

Drucilla Cornell

Thomas R. Flynn

Kevin Hart

Richard Kearney

Jean-Luc Marion

Adriaan Peperzak

Thomas Sheehan

Hent de Vries

Merold Westphal

Edith Wyschogrod

Michael Zimmerman

John D. Caputo, *series editor*

PERSPECTIVES IN
CONTINENTAL
PHILOSOPHY

CATHERINE MALABOU

What Should We Do with Our Brain?

Foreword by Marc Jeannerod
Translated by Sebastian Rand

FORDHAM UNIVERSITY PRESS
New York ■ 2008

Copyright © 2008 Fordham University Press

All rights reserved. No part of this publication may be reproduced, stored in a retrieval system, or transmitted in any form or by any means—electronic, mechanical, photocopy, recording, or any other—except for brief quotations in printed reviews, without the prior permission of the publisher.

What Should We Do with Our Brain? was first published in French as *Que faire de notre cerveau?* by Catherine Malabou © Bayard, 2004.

This work has been published with the assistance of the French Ministry of Culture—National Center for the Book.

Ouvrage publié avec le concours du Ministère français chargé de la culture—Centre National du Livre.

Library of Congress Cataloging-in-Publication Data is available.

Printed in the United States of America
10 09 08   5 4 3 2 1
First edition

Contents

Translator's Note

I would like to thank Catherine Malabou for her support and patience, Helen Tartar at Fordham Press for her enthusiasm for this project, and Sam Weber for giving me the opportunity to meet and work with the author. Thanks also to Alan Bass, Elana Commisso, Jason Leddington, and Richard Rand for timely and extensive comments on earlier drafts, to Paul Katz and Eddy Nahmias of Georgia State University for terminological assistance, and to Heather Cameron of the National Institutes of Health for providing me with an English copy of her essay quoted here.

Foreword

MARC JEANNEROD

The brain has always been described by means of technological metaphors: as an apparatus relaying excitation with
the precision of a mirror reflecting light, as a hydraulic
pump driving animal spirits into the muscles, as a central
telephone exchange connecting or cutting off communication; in the digital age, as a computer running its programs.
These metaphors, as Catherine Malabou remarks, proceed
from a centralizing concept of the brain seen as a machine
that works from the top down, that orders movement, con-
trols behavior, and brings about a unity of mind, consciousness, and man. In an earlier day, this centralizing and
unifying vision truly represented an ideal of governance:
one sole leader, one sole head commanding and organizing
everything. We recall the difficulties faced during the Napo-
leonic era by Franz Josef Gall. His system, subdividing the
mind into faculties distributed among different areas of the

brain, was seen by the ruling powers as a threat to the unity and stability of the state.

Brain plasticity shatters this concept. The machine learns, differentiates itself, reconstructs itself. Briefly put, it privileges the event over the law. Omnipresent plasticity changes our view of the brain and its functioning. But Malabou goes further, seeking to show that the transition from a wired brain to a plastic brain is really the transition from a "brain-machine" to a "brain-world." According to her, this change in perspective would affect not merely the model of cerebral functioning but also the concept we forge of ourselves and our social organization. The new model of the brain progressively elucidated by modern neuroscience emerges in a particular context: it co-occurs with a radical modification of the economic and social environment. The look of capitalism has changed, passing from a planned system, managed from above and overseen by a central authority, to an auto-organization at once dynamic, multipolar, and adaptive to circumstance. This new model of organization clearly suggests an analogy with cerebral reality: "Like neuronal cohesion, contemporary corporate economic and social organization is not of a central or centralizing type but rests on a plurality of mobile and atomistic centers, deployed according to a connectionist model" (p. 42). Might we have a "neo-liberal" brain that would impose its model on our socioeconomic organization? Or, inversely, might the global economy's upheaval generate a conceptual change that would affect, by contagion, our view of the way the brain functions?

The analogy between the functioning of a modern business enterprise and a brain is truly striking: in both cases, decision-making centers are distributed, and networks undo and recompose themselves according to tasks to be accomplished and goals to be realized. But this analogy has limits: in the brain, delocalization is not total, or, rather, localization is not fortuitous. It depends upon a connective organization emplaced at the embryonic phase. Each region, of the cortex in particular, therefore possesses a de facto specialization, determined by its connections and its position in the information-processing line. Thus the pertinent model would have a character more subsidiary than pluricompetent. Pathological focal lesions show that the alteration of a particular function (language processing, for example) can leave others intact (music processing). Similarly, the reuse of an intact zone to supplement a lesional zone works in very limited ways. Restoring a damaged network through a substitutive operation comes at the price of despecialization and performance loss. In the case of a systemic illness giving rise to depression, despecialization manifests itself as disaffiliation, a rupture in the social bond, placing the individual outside the network. Malabou rightly draws a parallel between illnesses of social connection, such as depression, and neurodegenerative illnesses, such as Alzheimer's dementia. Both cases involve a disconnection from the rest of the network. In one case, reconnection is possible; in the other, it is not. This is how dementia appears as the countermodel of plasticity: the irreparable loss of connections at the core of the cerebral network entails a definitive disconnection from the social network.

The analogy between cerebral organization and socioeconomic organization should thus, at the very least, lead us to an awareness of the relation between the subject and his brain. Abandoning the thesis of a rigid, predetermined, directing organ for the thesis of a supple, adaptable, plastic organ permits the political emancipation of the brain, the transition from a "soviet" to a "liberal" brain. But what is the consequence of this conceptual reversal for each individual? The brain itself has not changed. Humans in the

Middle Ages, the Industrial Revolution, and the liberal revolution all have had the same brain, with the same capacities for learning and adaptation. What changes is the organization of society, the outcome of organizational forces and macroscopic interactions over which the brain has little influence. Thus the problem is, rather, that of understanding how an individual brain can respond to the challenges of its social environment.

Malabou positions her book at the center of this questioning. For her, the brain makes possible the fundamental organic coherence of our personality, our self. The self is the result, the reflection, of the ordered functioning of the neuronal networks comprising the brain. This, in the end, guarantees the internal coherence of our representations. By guaranteeing the synthesis of the neuronal and the mental, the brain regulates the homeostasis of our mental states, just as, on another level, we have a regulation of the internal milieu maintaining the unity of the organism. We clearly have no consciousness of the plastic mechanisms forming our personality and guaranteeing its continuity. Yet by trying to become conscious of them we may, Malabou proposes, acquire a new freedom, that of imposing our own organization on the world rather than submitting to the influences of a milieu. Plasticity, in effect, is not flexibility. Let us not forget that plasticity is a mechanism for adapting, while flexibility is a mechanism for submitting. Adapting is not submitting, and, in this sense, plasticity ought not to serve as an alibi for submitting to the new world order being dreamed up by capitalism. "Not to replicate the caricature of the world: this is what we should do with our brain" (p. 78). To be conscious of the plasticity of one's brain is to give oneself the means to say no.

Acknowledgments

I would like to thank Professors François Ansermet and Pierre Magistretti, who, by inviting me to speak in Lausanne in the context of the *Semaine du Cerveau* ["Brain Week"]—an annual meeting in Switzerland with a large attendance—allowed me to meet, both at the seminar in child psychiatry and at the Centre Hospitalier Universitaire Vaudois, an entirely different audience than is usually mine. I thank Béatrice Bouniol for encouraging me to develop my hypotheses by inviting me into her series Le temps d'une question, in which the original French edition of this book appeared.

What Should We Do with Our Brain?

Introduction

Plasticity and Flexibility—For a Consciousness of the Brain

The brain is a work, and we do not know it. We are its subjects—authors and products at once—and we do not know it. "Humans make their own history, but they do not know that they make it," says Marx, intending thereby to awaken a consciousness of historicity. In a certain way, such words apply precisely to our context and object: "Humans make their own brain, but they do not know that they make it." It is not a question of effecting a tricky rhetorical move by corrupting this fine phrase for the benefit of our analysis or of acceding to the facility of a formal analogy. Quite the contrary, the bond between brain and history—concepts long taken to be antithetical—is now established with certainty.

The structural bond here is so deep that in a certain sense it defines an *identity*. It's not just that the brain has a history—which is sometimes confused with that of its constitution as an object of the sciences—but that it *is* a history.

In fact, today we can say that there exists a constitutive historicity of the brain. The aim of this book is precisely to awaken a consciousness of this historicity. It is no longer important to ask whether brain and consciousness are one and the same thing—let us put aside this old and specious debate. Instead we must constitute this strange critical entity, at once philosophical, scientific, and political, that would be a *consciousness of the brain*. It is to the constitution of this new genre—open to everyone—that the question *What should we do with our brain?* invites us.

We have not yet assimilated the results of the revolutionary discoveries made over the last fifty years in neurosciences (that is to say, the group of disciplines that study the anatomy, physiology, and functioning of the central nervous system, or CNS[1]), results that contribute more every day to the refutation of erroneous yet nonetheless mysteriously persistent pictures we still have of the brain. Already in 1979, in the preface to his book *Neuronal Man*, Jean-Pierre Changeux declared that our knowledge in the field of neurosciences had experienced

> an expansion matched only by the growth of physics at the beginning of the century and molecular biology in the 1950s. The impact of the discovery of the synapse and its functions is comparable to that of the atom or DNA. A new world is emerging, and the time seems ripe to open this field of knowledge to a wider public than the specialists and, if possible, to share the researchers' enthusiasm.[2]

But this communication, this opening to the public at large, this sharing of enthusiasm, never took place. Twenty-five years later, the assessment remains the same: "The impasse on the topic of the brain is, with few exceptions, total."[3] Even if many things have changed, even if the neurosciences have become "cutting-edge" disciplines, even if medical imaging has made palpable progress, even if the "cognitive sciences" have attained the status of autonomous disciplines,[4] and even if the number of articles about the CNS in the mainstream press has multiplied, neuronal man still has no consciousness.

In this sense, we are still foreign to ourselves, at the threshold of this "new world," which we fail to realize makes up our very intimacy itself. "We" have no idea who "we" are, no idea what is inside "us." Of course, we have all heard people talk about neurons, synapses, connections, networks, different types of memory. Everyone knows about neurodegenerative disorders, such as Alzheimer's or Parkinson's. Many of us have seen, in hospitals, the output screen of a functional neural-imaging machine. Some of us know that today it is possible, thanks to new MRI and PET technologies,[5] to observe the brain *in vivo*, in real time. Everyone says that psychoanalysis is losing importance, and everyone hears talk, right or wrong, about how the only effective cure for nervous depression is the chemical kind. We all know about MAOIs or SSRIs;[6] we are vaguely familiar with the words *serotonin*, *noradrenalin*, and *neurotransmitter*, and we all know about the neuronal origin of tobacco or drug addiction. We know that it is now possible to successfully transplant a hand and that the brain can reconstitute its bodily schema to include foreign members. We have heard about a certain ability on the part of the nervous system to repair, at least partially, some of the damage it sustains. The word *resilience* is not unknown to us.[7]

The problem is that we do not see the link that unifies all these phenomena, names, and situations, which are here purposefully listed completely at random and appear to have nothing in common with one another. This link exists nonetheless and is tied to the activity of the brain, to its manner of developing itself, of working, of *doing*. This link is tied to its meaning as a *work*, our work, and as history, our history, and as a singular destiny, our destiny.

The work proper to the brain that engages with history and individual experience has a name: *plasticity*. What we have called the constitutive historicity of the brain is really nothing other than its plasticity. *The plasticity of the CNS, nervous plasticity, neuronal plasticity, synaptic plasticity*—we run into this word in every neurology department of every medical school and of every university hospital, in the name of every neuroscientific research team[8]—it jumps out at us, in its many occurrences, every time we look under the word *brain* in the library. It constitutes the name of a specific discipline in scientific journals.[9] This frequency and omnipresence are not at all contingent. In fact, plasticity is the dominant concept of the neurosciences. Today it constitutes their common point of interest, their dominant motif, and their privileged operating model, to the extent that it allows them to think about and describe the brain as at once an unprecedented dynamic, structure, and organization.

Our brain is plastic, and we do not know it. We are completely ignorant of this dynamic, this organization, and this structure. We continue to believe in the " 'rigidity' of an entirely genetically determined brain,"[10] about which it is obviously completely in vain to ask: *What should we do with this*? Even the very word *brain* frightens us: we don't understand anything about it—all these phenomena, all these folds, ridges, valleys, localizations, this jargon that describes (we imagine) a series of fixed, indeed genetically programmed, entities, without any suppleness, without any improvisational ability. We don't understand this organization, which gives rise to so many unsettling metaphors in the register of command and government: a controller that sends orders down from on high, a central telephone exchange, a computer . . . all of this cybernetic frigidity, which only serves to alienate us from consciousness,[11] itself the only sign of life and liberty in a domain of implacable organic necessity, where movement and grace seem to be reduced to mere reflex.

Meanwhile, plasticity directly contradicts rigidity. It is its exact antonym. In ordinary speech, it designates suppleness, a faculty for adaptation, the ability to evolve. According to its etymology—from the Greek *plassein*, to mold —the word *plasticity* has two basic senses: it means at once the capacity to *receive form* (clay is called "plastic," for example) and the capacity to *give form* (as in the plastic arts or in plastic surgery). Talking about the plasticity of the brain thus amounts to thinking of the brain as something modifiable, "formable," and formative at the same time. Brain plasticity operates, as we shall see, on three levels: (1) the modeling of neuronal connections (developmental plasticity in the embryo and the child); (2) the modification of neuronal connections (the plasticity of synaptic modulation throughout life); and (3) the capacity for repair (post-lesional plasticity). "Plasticity in the nervous system means an alteration in structure or function brought about by development, experience, or injury."[12]

But it must be remarked that plasticity is also the capacity to annihilate the very form it is able to receive or create. We should not forget that *plastique*, from which we get the words *plastiquage* and *plastiquer*,[13] is an explosive substance made of nitroglycerine and nitrocellulose, capable of causing violent explosions. We thus note that plasticity is situated between two extremes: on the one side the sensible image of taking form (sculpture or plastic objects), and on the other side that of the annihilation of all form (explosion).

The word *plasticity* thus unfolds its meaning between sculptural molding and deflagration, which is to say explosion. From this perspective, to talk about the plasticity of the brain means to see in it not only the creator and receiver of form but also an agency of disobedience to every constituted form, a refusal to submit to a model.

Let us dwell for a moment on the modeling of neuronal connections, made possible by our individual experience, skills, and life habits, by the power of *impression* of existence in general. We can now see that the plasticity of the brain, understood in this sense, corresponds well to the possibility of fashioning by memory, to the capacity to shape a history. While the central nervous system's power for change is particularly clear during the developmental stage, we know for certain that the ability to learn, to acquire new skills and new memories, is maintained throughout life. And this is true *in a different way from one individual to the next*. The capacity of each to receive and to create his or her own form does not depend on any pre-established form; the original model or standard is, in a way, progressively erased.

Synaptic efficacy grows or declines under the impact of strictly individual experience. The synapse—from the Greek *sunapsis*, "liaison, juncture"—is the region of contact or connection between two neurons. The neuron, an elementary unit of nervous tissue, can be divided into three parts: the cellular body (protoplasm), the dendrites, and the axon, which are its extensions. It is by means of these extensions that connections (synapses) are established between two neurons. Dendrites, along with the cellular body, constitute what we call the *postsynaptic* side of the neuron. (This is where connections coming from "upstream" neurons arrive.) The axon constitutes the *presynaptic* part of the neuron: its endpoints are in contact with other "downstream" neurons.[14] Marc Jeannerod explains:

If a synapse belongs to a circuit in frequent use, it tends to grow in volume, its permeability increases, and its efficacy increases. Inversely, a little-used synapse tends to become less efficacious. The theory of synaptic efficacy thus allows us to explain the gradual molding of a brain under the influence of individual experience, to the point of making it possible for us, in principle, to account for the individual characteristics and particularities of each brain. We are dealing here with a mechanism of individuation that makes each brain a unique object despite its adherence to a common model.[15]

In this sense—we know this by now—the brain of a pianist is not strictly identical to that of a mathematician, a mechanic, or a graphic artist. But it is obviously not just a person's "trade" or "specialty" that matters here. The entire identity of the individual is in play: her past, her surroundings, her encounters, her activities; in a word, the ability that our brain—that every brain—has to adapt itself, to include modifications, to receive shocks, and to create anew on the basis of this very reception. It is precisely because—contrary to what we normally think—the brain is not already made that we must ask what we should do with it, what we should do with this plasticity that makes us, precisely in the sense of a work: sculpture, modeling, architecture. What should we do with this plastic organic art? It is already known that "synaptic plasticity, continuing throughout learning, throughout development as well as adulthood, sculpts each of our brains. Education, experience, and training make each brain a unique work."[16] What should we do with all this potential within us? What should we do with this genetically free field? What should we do with this idea of a truly *living* brain (modification of synaptic efficacy, as we will see, is already implicated in the most elementary level of animal life, and thus appears today to be one of the fundamental characteristics of living beings), a fragile brain, which depends on us as much as we depend on it—the dizzying reciprocity of reception, donation, and suspension of form that outlines the new structure of consciousness?

We can now understand why Jean-Pierre Changeux claims that the "discovery of the synapse and its functions" was as revolutionary as the discovery of DNA: the former brings to the latter a level of precision and a modification so significant that it seems almost to contradict it. Brain plasticity constitutes a possible margin of improvisation with regard to genetic necessity. Today it is no longer chance versus necessity, but chance, necessity, and plasticity—which is neither the one nor the other. "We know," says Changeux, "that the power of genes assures the perpetuation of broad traits of [cerebral] organization, such as the form of the brain and of its circumvolutions, the organization of its areas and the general architecture of cerebral tissue. . . . But considerable variability . . . remains despite the genes' power."[17] If neuronal function is an event or should bring about events, this is so precisely because it is itself able to create events, to eventualize [*événementialiser*] the program and thus, in a certain sense, to deprogram it.

We are living at the hour of neuronal liberation, and we do not know it. An agency within us gives sense to the code, and we do not know it. The difference between the brain and psychism is shrinking considerably, and we do not know it. "We" end up coinciding completely with "our brain"—because our brain is us, the intimate form of a "proto-self,"[18] a sort of organic personality—and we do not know it. Humans make their own brain, but they do not know they are doing so.

But why? Why do they not know it? Why do we persist in our belief that the brain is purely and simply a "machine," a program without promise? Why are we ignorant of our own plasticity? It is not because of a lack of information; exoteric books on the subject of brain plasticity abound. It is not because of a problem with popularization; we can talk in a very simple way about this plasticity, and that is precisely what this book is going to do. It is not a question of acquaintance but a question, once again, of consciousness. What must we be conscious of (and not merely acquainted with) concerning brain plasticity? What is the nature of its meaning?

We will respond, without playing on words, by saying that the consciousness we want to raise on the subject of plasticity has to do with its power to naturalize consciousness and meaning. Clearly, if we are not conscious of plasticity this is because, in accordance with a merely apparent paradox, it is in fact so familiar to us that we do not even see it; we do not note its presence, like an environment in which we maintain ourselves and evolve without paying attention to it. It has become the form of our world. As Luc Boltanski and Eve Chiapello note in their remarkable work *The New Spirit of Capitalism*, neuronal functioning and social functioning interdetermine each other and mutually give each other form (here again the power of plasticity), to the point where it is no longer possible to distinguish them. As though neuronal function were confounded with the natural operation of the world, as though neuronal plasticity anchored biologically—and thereby justified—a certain type of political and social organization. This is precisely what is meant by a "naturalization effect." The authors declare that we live in a "connectionist world with the coherence and immediacy of something natural." But this "naturalization effect is especially powerful in those disciplines which, aiming to connect biology and society, derive the social bond from implantation in the order of living organisms, or construct their representation of society on the basis of a physiological metaphor—not, as in the old organicism, cellular differentiation, but much more today on the metaphor of the neuron with its networks and flows."[19]

Humans make their own brain but they do not know that they do so. We are entirely ignorant of brain plasticity. Yet we are not at all ignorant of a certain kind of organization of labor—part-time jobs, temporary contracts, the demand for absolute mobility and adaptability, the demand for creativity . . . The brain is our work, and we do not know it. Yet we know very well that we live in a reticular society. We have understood that to survive today means to be connected to a network, to be capable of modulating one's efficacy. We know very well that every loss of suppleness means rejection, pure and simple. Is the difference really all that great between the picture we have of an unemployed person about to be kicked off the dole and the picture we have of someone suffering from Alzheimer's? We know already that individuals construct their lives as works, that it is each individual's responsibility to know what he should do with himself, and that for this he ought not be rigid. There is thus no need, in a certain sense, to be acquainted with the results of current discoveries in the neurosciences in order to have an immediate, daily experience of the *neuronal form of political and social functioning*, a form that today deeply coincides with the current face of capitalism.

The reference to Marx at the beginning of our analysis takes on its full importance here. In asking the question *What should we do with our brain?* we don't merely want to present the reader with some clarifications about cerebral functioning. Playing on the title of a well-known work by Daniel Dennett, we are not seeking to explain or explicate consciousness, but to *implicate* it.[20] To implicate consciousness, to ask what we should do with our brain, means, starting from these clarifications, to attempt to develop a critique of what we will call *neuronal ideology*. It is thus not just a matter of uncovering, in the name of brain plasticity, a certain freedom of the brain but rather, starting from as precise a study as possible of the functioning of this plasticity, to free this freedom, to disengage it from a certain number of ideological presuppositions that implicitly govern the entire neuroscientific field and, by a mirror effect, the entire field of politics—and in this way to rescue philosophy from its irresponsible torpor. Philosophers, excepting "cognitive scientists," are not sufficiently interested in the problem, mostly misunderstand the cognitive sciences, and, in the end, are simply ignorant of the results of recent research on the brain. So they miss the ideological stakes as well.

But *What should we do with our brain?* is not a question reserved for philosophers, for scientists, or for politicians—it is a question for everyone. It should allow us to understand why, given that the brain is plastic, free, we are still always and everywhere "in chains"; why, given that the activity of the central nervous system, as it is revealed today in the light of scientific discovery, presents reflection with what is doubtless a completely new conception of transformation, we nonetheless have the feeling that nothing is transformed; and why, given that it is clear that there can no longer be any philosophical, political, or scientific approach to history that does not pass through a close analysis of the neuronal phenomenon, we nonetheless have the feeling that we lack a future, and we ask ourselves *What good is having a brain, indeed, what should we do with it?*

The guiding question of the present effort should thus be formulated: *What should we do so that consciousness of the brain does not purely and simply coincide with the spirit of capitalism?* We will formulate the following thesis: today, the true sense of plasticity is hidden, and we tend constantly to substitute for it its mistaken cognate, *flexibility*. The difference between these two terms appears insignificant. Nevertheless, flexibility is the ideological avatar of plasticity—at once its mask, its diversion, and its confiscation. We are entirely ignorant of plasticity but not at all of flexibility. In this sense, plasticity appears as the coming consciousness of flexibility. At first glance, the meanings of these two terms are the same. Under the heading "flexibility," the dictionary gives: "firstly, the character of that which is flexible, of that which is easily bent (elasticity, suppleness); secondly, the ability to change with ease in order to adapt oneself to the circumstances." The examples given to illustrate the second meaning are those that everybody knows: "flexibility on the job, of one's schedule (flex time, conversion), flexible factories." The problem is that these significations grasp only one of the semantic registers of plasticity: that of receiving form. To be flexible is to receive a form or impression, to be able to fold oneself, to take the fold, not to give it.[21] To be docile, to not explode. Indeed, what flexibility lacks is the resource of giving form, the power to create, to invent or even to erase an impression, the power to style. Flexibility is plasticity minus its genius.[22]

Humans make their own brains, and they do not know that they do so. Our brain is a work, and we do not know it. Our brain is plastic, and we do not know it. The reason for this is that most of the time flexibility superimposes itself on plasticity, even in the midst of scientific discourses that take themselves to be describing it entirely "objectively." The mistake in certain cognitivist discourses, for instance, is not that they reduce the mental to the neuronal or the mind to a biological entity. I am myself entirely materialist, and such affirmations do not shock me at all. The error is in thinking that neuronal man is simply a neuronal given and not also a political and ideological construction (including of the "neuronal" itself). One notes that many descriptions of plasticity are in fact unconscious justifications of a flexibility without limits. Sometimes it seems as though in nervous systems, from the aplysia[23] to the human, a faculty is deployed—a faculty described precisely in terms of synaptic plasticity—to fold, to render oneself docile vis-à-vis one's environment, in a word, to adapt to everything, to be ready for all adjustments. It is as though, under the pretext of describing synaptic plasticity, we were really looking to show that flexibility is inscribed in the brain, as though we knew more about what we could stand than about what we could create. That said, securing a true plasticity of the brain means insisting on knowing what it can do and not simply what it can tolerate. By the verb *to do* or *to make* [*faire*] we don't mean just "doing" math or piano but making its history, becoming the subject of its history, grasping the connection between the role of genetic nondeterminism at work in the constitution of the brain and the possibility of a social and political nondeterminism, in a word, a new freedom, which is to say: a new meaning of history.

Flexibility is a vague notion, without tradition, *without history*, while plasticity is a *concept*, which is to say: a form of quite precise meanings that bring together and structure particular cases. This concept has a long philosophical past, which has itself remained too long in the shadows. I do not intend to criticize anyone here, and my goal is not polemical. I would simply like to disentangle the notion and the concept, to make us stop taking the one for the other and conflating them, as I have intentionally done above, in speaking simultaneously of nervous depression, hand transplants, and lesion repair. I would like to do this in order to stigmatize the definitional magma in which, in the end, we all bathe, the author of this book along with everyone else. Speaking for myself, I would say that I have been interested for a long time in plasticity, whose genesis and whose meaning in the philosophical tradition I have, in previous efforts, attempted to elucidate and reconstitute.[24] The study of neuronal plasticity and cerebral functioning, and the reading of important texts by cognitive scientists dedicated to this functioning, have been much more than an enrichment for me: they have been a true test as well as a confirmation, a renewal, and a concretization of the philosophical meaning of plasticity. The critical epistemological exercise carried out in this book thus presents itself as an enterprise of rectification and sharpening of the usage of this concept.

But let us not forget that the question *What should we do with our brain?* is a question *for everyone*, that it seeks to give birth in everyone to the feeling of a new responsibility. The inquiry conducted here thus ought, beyond the critical imperatives just announced, to allow anyone who consents to follow its path to think new modalities of forming the self, under the name of "plasticity" and beyond the overly simplistic alternative between rigidity and flexibility. This means asking not "To what point are we flexible?" but rather "To what extent are we plastic?"

1

Plasticity's Fields of Action

Between Determination and Freedom

In mechanics, a material is called *plastic* if it cannot return to its initial form after undergoing a deformation.[1] "Plastic" in this sense is opposed to "elastic." Plastic material retains an imprint and thereby resists endless polymorphism. This is the case, for instance, with sculpted marble. Once the statue is finished, there is no possible return to the indeterminacy of the starting point. So plasticity designates solidity as much as suppleness, designates the definitive character of the imprint, of configuration, or of modification. According to this first limit or semantic extreme, plasticity, though not altogether assimilable to rigidity, marks a certain determination of form and imposes a (very strict) restriction on the capacity for deformation, re-formation, or explosion. We will see that this somewhat "closed" or restrained signification is essentially at work in the developmental plasticity of neuronal connections tied to the genetic determinism that presides over the constitution of every brain.

The second limit on the range of the concept of plasticity is marked, inversely, by an "open" or unrestrained definition. According to this second limit, plasticity designates a much more effective transformative ability. This involves, not an infinite modifiability—we have not yet come back around to polymorphism—but a possibility of *displacing* or *transforming* the mark or the imprint, of changing determination in some way. As an example of such a meaning, let us consider the properties of so-called "adult" stem cells (at work in the adult organism and thereby distinguished from "embryonic" stem cells). Adult stem cells are nonspecialized cells found in specialized tissues (the brain, bone marrow, blood, blood vessels, the retina, the liver, etc.). They renew themselves, and most of them specialize, in order to produce all the types of cells in their tissue of origin that normally die. This is how, for example, immature blood cells are made out of bone marrow stem cells. But while the majority of adult stem cells generate cells similar to those of the tissue they come from, it has been discovered that some of them (notably skin stem cells) can transform themselves into different types of cells (for example, nerve or muscle cells). One then says that they "transdifferentiate" themselves, that is, literally, that they change their difference.[2]

This capacity to differentiate and transdifferentiate themselves is called, precisely, stem-cell plasticity. In the first case—the capacity to differentiate themselves into cells of the same tissue—stem cells are called *multipotent*.[3] In the second case—the capacity to develop themselves into types of cells specific to other tissues—stem cells are called *pluripotent*.[4] Stem-cell plasticity—which allows us to conceive of a sort of range of differentiation between multipotence and pluripotence—is an extremely striking example, perhaps the very paradigm, of the "open" meaning of plasticity. According to this meaning, plasticity designates generally the ability to change one's destiny, to inflect one's trajectory, to navigate differently,[5] to reform one's form and not solely to constitute that form as in the "closed" meaning. This open meaning is essentially at work in the *plasticity of synaptic modulation*, as we will see when we study the interplay of the modification of synaptic connections and "secondary neurogenesis"—the renewal of neurons in the adult brain, starting, precisely, from stem cells.

Thus, with plasticity we are dealing with a concept that is not contradictory but graduated, because the very plasticity of its meaning situates it at the extremes of a formal necessity (the irreversible character of formation: determination) and of a remobilization of form (the capacity to form oneself otherwise, to displace, even to nullify determination: freedom). It is this complex, this synthesis, this semantic wealth, that we ought to keep in mind throughout our analysis.

The Three Plasticities

We will now look more closely at the biological phenomenon of brain plasticity according to its three major roles: developmental plasticity, modulational plasticity, and reparative plasticity.

Developmental Plasticity: The Formation of Neuronal Connections

What do we find in the brain? Billions of neurons (around twenty billion in humans) connected in a network of innumerable links, the synapses. "The human brain," says Changeux, "makes one think of a gigantic assembly of tens of billions of interlacing neuronal 'spider's webs' in which myriads of electrical impulses flash by, relayed from time to time by a rich array of chemical signals."[6] These "spider's webs," neuronal connections also called "arborizations," are constituted progressively over the course of an individual's development. We use the term *plasticity* precisely to characterize this neuronal genesis. The brain, in effect, forms itself. "The human infant is born with a brain weighing about 300 grams—20 percent of the weight of an adult brain. . . . One of the major features of the development of the human brain, then, is that it continues well after birth, for about 15 years."[7]

Everything begins with establishing connections and then multiplying them and making them more complex. The growth in mass of the brain coincides with the extension of axons and dendrites, the formation of synapses, and the development of myelin sheaths around the axons. This development is subject to strict genetic determinism. As Jeannerod notes, from the point of view of their genesis and their constitution, "all human brains resemble each other."[8] The connections that constitute the anatomy of the mature brain are obviously not the result of chance or of spontaneous arrangement; the migration of nerve cells and their adaptation to their targets are programmed. He continues:

> To take just one example, the fibers that come from the retina and transport visual information end their journey, in all individuals, in the visual part of the cortex—that is, in the occipital lobe, occupying the rear part of the brain; in all individuals, connections are established between this visual region and other regions situated in the parietal lobe and in the temporal lobe, and so forth. The adult brain therefore reflects the existence of a pre-established plan that

causes brain anatomy to be the same from one individual to the next.[9]

But if neuronal genesis corresponds to a "pre-established plan," why talk about *plasticity* in order to characterize this development? For two essential reasons, which, within the context of development, have to do with (1) establishing connections, a process we have just mentioned, and (2) modeling those connections (which ought not to be confused with the modulation of synaptic efficacy). In both cases, it is the execution of the genetic program that works in a plastic way. There is a sort of plastic art of the brain—hence the use of the term *plasticity* in this context. And it is here that the restrained or "closed" signification of the concept has to be taken into account: the sculpting of a determinate form.

In the course of the process of establishing connections, the sculptor's chisel is the phenomenon called "apoptosis" or "cell death." This death is a normal phenomenon. Again it corresponds to the execution of a genetic program, leading to the elimination of useless connections and to the progressive sculpting of the definitive form of the system by fitting nerve fibers to their targets. In the human brain, neuronal death begins at the end of gestation and continues after birth, for at least the first six months of life. It continues in adults at a much slower pace. "This neuronal 'sacrifice,'" writes Changeux, "is part of normal development; indeed, it constitutes one of its critical phases."[10] In an eloquently titled book, *The Sculpture of the Living*, Jean-Claude Ameisen insists that the brain, far from being, as was previously believed, an organ fully constituted at birth, simultaneously receives and gives itself form. "Cell death," he writes, "is . . . a tool allowing the embryo to work out its form in its becoming, by an eliminative procedure that allies it with sculpture."[11]

From this stage of development on, however, once the definitive form of the system has been sculpted, "genetic determination begins to slacken,"[12] explains Jeannerod. "After birth, the topographic network put in place during embryogenesis and stabilized by neuronal death and by the elimination of connections begins to function under the influence of external factors. This functioning brings with it a new phase of modeling of connections."[13] The role of the surroundings is therefore fundamental here. A great deal of the development of the human brain is accomplished in the open air, in contact with the stimuli of the world, which directly influence both the development and the volume of connections. The visual system, for example, is not entirely functional at birth. The synapses connecting fibers coming from the retina to neurons in the visual cortex are not yet entirely formed. Information received from outside activates the synapses and encourages maturation. In this sense, in the second phase of development one can speak of a modeling of synapses or a mechanism of synaptic plasticity—always tied, as we have seen, to a genetic program.

The genesis of the brain, through the two phases of establishing connections and their maturation under the influence of the surroundings, thus makes evident a certain plasticity in the execution of the genetic program. In both cases, the brain appears at once as something that gets formed—progressively sculpted, stabilized, and divided into different regions—and as something formative: little by little, to the extent that the volume of connections grows, the identity of an individual begins to outline itself. But the more time passes, the more this "first plasticity" loses its determinist rigor. The sculptor gradually begins to improvise. Bit by bit, the modeling becomes that which our own activity imprints on the connections: "our brain . . . modeled by our own activity, by our interactions with the external world, by the influences we have received in the course of our education, knows our history and our trajectory. From this intimacy is born a profound identity between the functioning of our brain and our conception of the world, an identity of views, one might say."[14]

In fact, this first type of plasticity is closely tied to the second, both because the influence of the surroundings gradually takes over from epigenetic sculpting and because it engages in a more and more precise activity. The restrained or "closed" meaning of plasticity very quickly runs into its "open" signification: the "freedom" in which determinacy and nondeterminism cross paths in an astonishing way. Indeed, we see that cerebral morphogenesis results not in the establishment of a rigid and definitively stable structure but rather in the formation of what we might call a *template*. This is then refined (sculpted) during development and, in a subtler but always powerful way, throughout life. The nervous activity of pre-established circuits thus takes over from apoptotic sculpting. Henceforth the environment of the brain qua organ (the modeling of connections) and its external environment (synaptic modulation by influence of the surroundings) play the role of morphogenic factors.

Modulational Plasticity: The Brain and Its History

At this point, we immediately encounter brain plasticity's second field of action: the modification of neuronal connections by means of the modulation of synaptic efficacy. Without a doubt, it is at this level that plasticity imposes itself with the greatest clarity and force in "opening" its meaning. In effect, there is a sort of neuronal creativity that depends on nothing but the individual's experience, his life, and his interactions with the surroundings. This "creativity" is not reserved solely for the human brain but is already at work in the most rudimentary nervous systems.

Such a plasticity, consisting in the fashioning of interconnections and in the modulation of synaptic efficacy, was first brought to light by the Canadian neurologist Donald Holding Hebb.[15] At the end of the 1930s, various experimental observations led him to abandon the concept of a rigid localization of memory circuits along the lines of the model of reflex circuits described by Pavlov. According to Hebb, we must postulate the existence of "plastic synapses" capable of adapting their transmission efficacy. Hebb formulated the hypothesis of neuronal circuits capable of self-organization, that is, of modifying their connections during the activity required for perception and learning. The synapse is the privileged locus where nerve activity can leave a trace that can displace itself, modify itself, and transform itself through repetition of a past function.[16]

The capacity of synapses to modulate their efficacy and to modify the force of their interconnections under the influence of experience works in a double sense. The efficacy of the synapse (its capacity to transmit signals from neuron to neuron) either rises, which is called "long-term potentiation" (LTP), or diminishes, which is "long-term depression" (LTD). This can be verified even in an animal like the aplysia. Its central nervous system is simple, composed of eight pairs of ganglia situated around its esophagus and one large abdominal ganglion. The aplysia has a small set of stereotypical behaviors, among them a number of protective maneuvers, such as retracting its siphon and its gills. But the intensity of its self-defense reflex is modulated by experience. Repeated innocuous stimulation of its mantle results in a diminution of the reflex (a habituation), which manifests as a decrease in the size of the retracting motion. This habituation is accompanied by a depression in synaptic activity correlated to the amount of neurotransmitter emitted at the level of the motor-sensory synapse.[17]

The phenomena of long-term depression and potentiation show up with much more clarity in the processes of adaptation, learning, and memory at work in birds. The black-headed titmouse, for example, stores food in caches and later retrieves it, practically infallibly. Researchers have been able to establish that the size of one of the brain regions (the hippocampus[18]) implicated in this process is greater in this bird than in others that do not stockpile their food. That is, species that practice such stockpiling have significantly larger hippocampuses than others. This change results from a growth in the number of new neurons, from a diminution in cell death (apoptosis), and from an increase in the connections between the neurons of the hippocampus. The hippocampus thus manifests a remarkable structural plasticity.[19]

Potentiation and depression are not just synaptic processes in which one or more stimuli induce immediate activations; they are also long-term modifications, capable of changing form (a change in the size of brain region, a variation in the permeability of a regularly activated region) and of undoing a trace in order to remake it differently (the lability of the mnemonic trace). Generally, some nerve networks increase their performance by "depressing" synapses involved in cognitive tasks that have led to errors in the course of motor-system education. This phenomenon shows up quite clearly in the human brain during all learning processes. In the course of learning to play the piano, for example, the mechanism for depressing entry signals corresponding to incorrect movements ("mistakes") makes possible the acquisition of the correct movements. In the case of potentiated connections, synapses enlarge their area of contact, their permeability rises, and nerve conductivity is more rapid. Inversely, a little-used or "depressed" synapse tends to perform less well. Neurons somehow *remember* stimulation. Everything happens as if there were no stabilization of memories except on the condition of a potential destabilization of the general landscape of memory.[20]

Long-term potentiation is therefore structurally tied to long-term depression,[21] and this tie constitutes the differentiating or transdifferentiating force of neuronal plasticity. By analogy with the process of becoming that stem cells undergo, one could claim that neuronal connections, because of their own plasticity, are always capable of *changing difference*, receiving or losing an imprint, or transforming their program.

The fact that synapses can see their efficacy reinforced or weakened as a function of experience thus allows us to assert that, even if all human brains resemble each other with respect to their anatomy, no two brains are identical with respect to their history. The phenomena of learning and memory show this directly. Repetition and habit play a considerable role, and this reveals that the response of a nervous circuit is never fixed. Plasticity thus adds the functions of artist and instructor in freedom and autonomy to its role as sculptor. In a certain sense, it is possible to assert that the synapses are the future reserves of the brain. They are not immobilized and do not constitute simple transmitters of nervous information but rather have the power to form or to reform this very information. "The efficacy of the synapses," declares Jeannerod, "varies with respect to the flux of information traversing them: during infancy and throughout life, each one of us is subject to a unique configuration of influences from our external surroundings, which resonates in the form and the functioning of the brain's networks."[22]

This allows us to put back into question the old dogma that the adult brain steadily loses its plasticity, the dogma that the brain can of course acquire new information but can know no great change in its capacity to learn, its memory function, or its global structures except in the direction of decline or degeneracy. On the contrary, we see that there exists an ongoing reworking of neuronal morphology.

Reparative Plasticity: The Brain and Its Regeneration

This point leads to our treatment of plasticity's third field of action: repair. Two distinct processes fall under the heading of reparative plasticity: neuronal renewal, or secondary neurogenesis, and the brain's capacity to compensate for losses caused by lesions.

What are we to understand by "neuronal renewal" or "secondary neurogenesis"? According to what we have just said, it would seem that a primary plasticity—morphogenic —is followed by a modulational plasticity that modifies synaptic efficacy but does not affect the anatomical stability of the brain, as though this plasticity somehow operated inside a closed system. "Certain scientists," declares Heather Cameron, "still cling to a very rigorous form of the hypothesis of a stable brain, according to which there is no anatomical plasticity in the adult brain, and especially not in the cortex; they hold that the functional plasticity underlying learning mechanisms requires only modifications in the force of the synapses, produced by a modification in the receptors or in the intercellular environment of the neuron at the molecular level."[23] But this dogma of the stable brain is not quite right. In fact, she continues, "we know already that certain neurons in regions important to the learning process renew themselves continuously—which constitutes a relatively important anatomical modification." Even if the role of stem cells in the adult brain and their localization still remain poorly known, even if it is probable that secondary neurogenesis does not affect all regions of the brain, a renewal of nerve cells in adulthood exists all the same, a renewal that, in opening untapped perspectives on brain repair, modifies the way in which we must view the functioning of the brain.

A recent study of the neocortex in primates has produced evidence of new neurons in three regions of the associative cortex: the prefrontal region, the inferior temporal region, and the posterior parietal region. "This result is particularly interesting because the associative cortex plays an important role in high-level cognitive functions, while the striate cortex [in which no renewal is observed] participates in the handling of information with a visual origin. This difference makes one think that neurogenesis could play a key role in essentially plastic functions, while it would be pointless for low-level functions like the handling of sensory data, which functions are generally stable throughout life."[24]

The production of new neurons therefore does not simply serve to replace cells that have died; it participates in modulational plasticity and, in doing so, opens the concept of plasticity slightly more, just far enough to unsettle the concept of stability. Once more: the statue is alive, the program quickens itself; right where we have so often believed we would find only mechanism, we find a complex entanglement of different types of plasticity, which contradicts the ordinary representation of the brain as machine. Alain Prochaintz affirms:

> One of the major characteristics of the nervous system is, without a doubt, its plasticity. The brain cannot be considered to be a network of permanently established cables, with cerebral aging being the result of an

increasing number of units in this circuit becoming disconnected from the network and going out of operation. Although this has not positively been demonstrated except in a few experimental models, we can assume that every day new fibers are growing, synapses are becoming undone, and new ones are being formed. These changes in the neuronal . . . landscape mark our capacity for adaptation, our capacity for learning and improvement, which continue until an advanced age, and in fact until death.[25]

In an article entitled "The Curious Partition of New Neurons,"[26] researchers assert that "in light of observations of secondary neurogenesis, it appears clear that the adaptive capacities of the nervous systems of birds and adult mammals are not solely the result of variations in synaptic connections. They are dependent on the production or the renewal of neurons in some very precise regions—regions that have the common characteristic of having functions tied to learning and/or memory. In this context, secondary neurogenesis also seems to permit a subject's personal experience routinely to leave an imprint on the core of neuronal networks, in the form of regular morphological and functional reworking. Adult neurogenesis, being the final mechanism of plasticity and one strongly controlled by a subject's personal experience and environmental interactions, very likely constitutes an additional mechanism of individuation—with the major difference that it is operational throughout life."[27]

The idea of cellular renewal, repair, and resourcefulness as auxiliaries of synaptic plasticity brings to light the power of *healing*—treatment, scarring, compensation, regeneration, and the capacity of the brain to build natural prostheses. The plastic art of the brain gives birth to a statue capable of self-repair. We know full well that the functioning of the brain can be disturbed by numerous pathological causes, the best-known being cranial trauma, strokes, encephalitis, and neurodegenerative disorders (Parkinson's, Alzheimer's). But the nervous system always demonstrates plasticity after such handicaps or lesions, whether or not these efforts are crowned with success: the affected structures or functions try to modify themselves so as to compensate for the new deficit or form a new and abnormal organizational schema that restores normalcy.

Reparative plasticity obviously does not make up for every deficit; we know that certain neuronal lesions are irreversible. But at the beginning, in the brain, there is always a more or less successful, more or less efficacious, more or less durable attempt to reorganize the affected function. Jeannerod takes as an example the phenomenon of

> the paralysis of the left arm provoked by a lesion on the right side of the motor region of the cortex following a stroke. At the start, all movement is impossible; the arm is immobile and flaccid. After a certain period, the muscular force returns, and elbow and wrist movements reappear. How is this possible if the neurons responsible for controlling these movements have been destroyed? . . . Functional neural imaging is very useful here: it shows us that during the patient's efforts to move the paralyzed arm, the left side of the motor region of the cortex, spared by the lesion, is activated. The patient, by himself or through rehabilitation, has learned to use nerve pathways that would not be there in the normal state. This reorganization of motor function testifies once more to the plasticity of brain mechanisms.[28]

Another example is what happens at the onset of Alzheimer's disease. The encroaching amnesia is compensated for in part by a capacity to recuperate stored information. The deactivation of certain regions (the region of the hippocampus) is balanced by a metabolic activation of other regions (the frontal regions). Thus after certain circuits are affected, there is a modification in strategies for handling information, a modification that again attests to the functional plasticity of the brain.[29]

There are, therefore, functions for postlesional reorganization. These phenomena can also be observed in certain transplants. In January 2000, a team from Edouard Harriot Hospital in Lyon performed the first human double hand transplant on Denis Chatelier, thirty-three years old, whose hands had been amputated four years earlier following an accidental explosion. The question was: Even if one manages to re-establish a precise anatomical continuity between the donor's hands and the recipient's forearms, can one attain the same continuity on the psychological and neurological level? The Chatelier case showed that one can. His phantom pains disappeared, and the motor progress he made allows us to assert that his brain succeeded in integrating his transplanted hands. "When the motor cortex reorganizes itself, the synapses modify themselves. They change their relative influence, their 'weight' in the local functioning of the network of neurons. . . . After the transplant, such a change in neuronal connections could come to restore the representation of the hand."[30] Yet more proof of our brain's striking capacity for adaptation.

Are We Free to Be High Performing?

We can see it now: there are not just *one* but *many* plasticities of cerebral functioning. The interaction of these plastic modalities sketches an organization that does not at all correspond to traditional representations of the brain as a machine without autonomy, without suppleness, without becoming—representations that today have become true "epistemological obstacles." It is urgent that we affirm, against these representations, which no longer represent anything at all, that our brain is in part essentially *what we do with it*. Individual experience opens up, in the program itself, a dimension usually taken to be the very antithesis of the notion of a program: the historical dimension. Plasticity, between determinism and freedom, designates all the types of transformation deployed between the closed meaning of plasticity (the definitive character of form) and its open meaning (the malleability of form). It does this to such a degree that cerebral systems today appear as self-sculpted structures that, without being elastic or polymorphic, still tolerate constant self-reworking, differences in destiny, and the fashioning of a singular identity.

The question that inevitably poses itself is thus: How can we know how to respond in a plastic manner to the plasticity of the brain? If the brain is the biological organ determined to make supple its own biological determinations, if the brain is in some way a self-cultivating organ, which culture would correspond to it, which culture could no longer be a culture of biological determinism, could no longer be, in other words, a culture against nature? Which culture is the culture of neuronal liberation? Which world? Which society?

The concept of plasticity has an aesthetic dimension (sculpture, malleability), just as much as an ethical one (solicitude, treatment, help, repair, rescue) and a political one (responsibility in the double movement of the receiving and the giving of form). It is therefore inevitable that at the horizon of the objective descriptions of brain plasticity stand questions concerning social life and being together. To expedite matters, let us reduce these to one option: Does brain plasticity, taken as a model, allow us to think a multiplicity of interactions in which the participants exercise transformative effects on one another through the demands of recognition, of non-domination, and of liberty? Or must we claim, on the contrary, that, between determinism and polyvalence, brain plasticity constitutes the biological justification of a type of economic, political, and social organization in which all that matters is the result of action as such: efficacy, adaptability—unfailing flexibility?

2

The Central Power in Crisis

These questions, of course, concern the governing and command functions immediately attributed to the brain. It is because in each individual the brain constitutes the controlling authority par excellence that all the descriptions we can give of it always participate, in one way or another, in political analysis. We can thus affirm that there is no scientific study of the modalities of cerebral power that does not by the same token—implicitly and usually unconsciously—adopt a stance with respect to the contemporary power of the very study within which it operates. There is today an exact correlation between descriptions of brain functioning and the political understanding of commanding.

What is the main transition point between the neuronal and the political? The foregoing descriptions of brain plasticity allow us to respond immediately: it has to do with putting centrality back into question. The metaphor of the central organ has definitively been surpassed, even if it continues to impose itself as an epistemological and ideological obstacle. This crisis of centrality rests on a delocalization and a reticular suppleness in the structures of command. In the same way that neuronal connections are supple and do not obey a centralized or even truly hierarchized system, political and economic power displays an organizational suppleness in which the center also appears to have disappeared. The biological and the social mirror in each other this new figure of command.

The End of the "Machine Brain"

The Central Telephone Exchange and the Computer

This new figure explains the fruitlessness of the well-known technological metaphors that have been used to characterize brain functioning. Essentially, these are mechanical metaphors, which turn the brain—as they do machines—into a *control center*. The two most famous, today put back into question by the discovery of plasticity, are the "central telephone exchange brain" and the "computer brain." The two assimilate the brain to a center and its organization to a process of centralization.

In *Matter and Memory*, Henri Bergson develops a famous analogy between the brain and a central telephone exchange. For Bergson, the role of the brain is limited to that of centralizing information. The brain does not produce representations; it contents itself with collecting them, sending them up the line, bringing them down the line, and circulating them: "in our opinion . . . the brain is no more than a kind of central telephonic exchange: its office is to

GARDNER HARVEY LIBRARY
Miami University-Middletown
Middletown, Ohio

allow communication or to delay it. It adds nothing to what it receives . . . but it really constitutes a center."[1] Bergson seeks to determine the role the brain plays in action: like a central telephone exchange, it puts things in relation but does not intervene in the relation itself. In this way, having no power either to create or to improvise, it does nothing beyond passing on messages. Jeannerod, commenting on these propositions of *Matter and Memory*, explains:

> The brain relates nervous excitation coming from the periphery to the motor mechanism. In the case of reflex motion, the excitation is propagated directly to the motor mechanisms of the medulla, and action is immediate. In the case of a more complex action, related to a perception, it takes a detour through the sensory cells of the cerebral cortex before descending again to the medulla. What has it gained by this detour? Certainly not the power to transform itself into a representation, which is useless, according to Bergson, but only the fact of being able to be connected, by the cells of the motor cortex, to the set of motor mechanisms of the medulla and thereby the power to choose its effect freely.[2]

As fascinating as it may be, this metaphor of the central telephone exchange is today outdated because it completely fails to capture plasticity and does not take into account synaptic and neuronal vitality.[3]

The cybernetic metaphor has also had its day. One of the subsections of Jeannerod's book *The Nature of Mind* is entitled "The Comparison Between Brain and Computer Is Not Pertinent."[4] This comparison dates to the fifties and reigned until the end of the eighties. It allowed research in Artificial Intelligence to make considerable progress.[5] The common trait of the brain and the computer is inarguably the notion of the program: the brain then would have a central programming function. Very simply, the analogy between the cybernetic domain and the cerebral domain rests on the idea that thinking amounts to calculating, and calculating to programming. The computer and the brain would in the end both be "thinking machines," that is, physico-mathematical structures endowed with the property of manipulating symbols. The discovery of the plasticity of brain functioning has rendered such a comparison moot. Plasticity invalidates not the analytical or explicative value of the mechanical paradigm in itself—a paradigm that is, to a certain extent, indispensable for thinking about brain function—but rather the *central* function habitually associated with the computer and its programs. Opposed to the rigidity, the fixity, the anonymity of the control center is the model of a suppleness that implies a certain margin of improvisation, of creation, of the aleatory. As Jeannerod says: "the activity of the nervous system can be better represented as the outline of a multidimensional map than as a sequence of symbols."[6] The representation of the center collapses into the network.

The interaction of the brain with its surroundings instead acts as a commanding authority, whose unknown form and location disrupt the traditional geography of government. The functional plasticity of the brain deconstructs its function as the central organ and generates the image of a fluid process, somehow present everywhere and nowhere, which places the outside and the inside in contact by developing an internal principle of cooperation, assistance, and repair, and an external principle of adaptation and evolution. "The brain would thus no longer be an organ that transfers the commands of the mind to the body, a kind of controller working from the top down, but rather a system that continuously proposes solutions compatible with our history and our needs."[7]

Gilles Deleuze, who is one of the rare philosophers to have taken an interest in neuroscientific research since the 1980s, goes so far as to talk of the brain as an "acentered system," "the effect of a break with the classical image" that has been formed of it.[8] Cerebral space is constituted by cuts, by voids, by gaps, and this prevents our taking it to be an integrative totality. In effect, neuronal tissue is discontinuous: "nerve circuits consist of neurons *juxtaposed* at the synapses. There is a 'break' between one neuron and the other."[9] Between two neurons, there is thus a caesura, and the synapse itself is "gapped." (One speaks, moreover, of "synaptic gaps.") Because of this, the interval or the cut plays a decisive role in cerebral organization. Nervous information must cross voids, and something aleatory thus introduces itself between the emission and the reception of a message, constituting the field of action of plasticity.

This specific distribution of information, which contradicts the idea of continuity, also disrupts the picture of vertical organization. The "discovery of a probabilistic or semifortuitous cerebral space, 'an uncertain system,'" according to Deleuze, implies the idea of a multiple, fragmentary organization, an ensemble of micro-powers more than the form of a central committee. In consequence, "our lived relation with the brain becomes more and more fragile, less and less 'Euclidean,' and goes through little cerebral deaths.[10] The brain becomes our problem or our illness, our passion, rather than our mastery, our solution or decision."[11] There thus exists a lived brain but, as I indicated at the start, this lived brain is not necessarily conscious. The proof of this is that the intimate feeling of cerebral fragility, constantly sustained by media images of neurodegenerative disorders, has not managed to usurp the dominant representation of a rigid centrality that is not pertinent even for describing machines.

I will not broach here the immense problem of the comparison between brain and machine in general. That would be another debate and another book. I would simply like to analyze the ideological cliché attached to the functioning of brains as much as to that of machines, the cliché of a centered and centralizing program that leaves no room for plasticity and entertains no relation with alterity. Why does this cliché, despite being undermined by scientific discoveries, have such endurance? Why does it prevent us from clearly thinking about and conceptualizing what, in effect, we *live*, what in many respects we make—our brains, which are, once more, our work, fashioned throughout a whole life within the intimate experience of the outside? Why doesn't the resolutely obsolete character of cybernetic metaphors, revealed by current research on brain plasticity, leap out at us more clearly, given that we live in a period of "weak" Artificial Intelligence?[12] And why do the same metaphors, the same clichés, equally prevent us from clearly thinking and conceptualizing what we live with our computers? Why do they still force us to hold onto a low-grade antitechnological discourse supported by the supposed omnipotence of the program-center?

Daniel C. Dennett's *Consciousness Explained* is one of the best books devoted to the problem of the comparison between brain and computer. He vindicates the foundations of the analogy (not the identity) between the two.[13] But in order to justify this foundation, he does not advance the arguments one would expect. In effect, Dennett presents the computer as itself a plastic organization, with multiple and supple levels of command. The comparison between brain and computer rests on this plasticity, which thus serves as analogon. "A computer," he writes, "has a basic *fixed* or *hard-wired* architecture but with huge amounts of plasticity thanks to the *memory*."[14] But how are we to characterize this plasticity? Like the brain, the machine Dennett describes is, against all expectation, "a virtuoso future-producer, a system that can think ahead, avoid ruts in its own activity, solve problems in advance of encountering them, and recognize entirely novel harbingers of good and ill."[15]

What we can take away from this analysis is an approach to the machine that thinks of it not as a control center but as an organ with multiple and adaptable structures—a future-producing organization, susceptible to an always-accruing functional differentiation, a machine somehow determined by the relation to alterity—a machine capable of privileging events over laws. It is not important here to determine whether such a machine exists, but simply to insist that this conception says out loud what we live deep inside, more precisely, that "computers are not 'number-crunching machines,'"[16] something we experience daily, and that plasticity perhaps designates nothing but the eventlike dimension of the mechanical.

The Adequation of Brain and World

Nevertheless, as we've said, the clichés of the center, of deterministic programming, and of blind mechanics endure. We persist in thinking of the brain as a centralized, rigidified, mechanical organization, and of the mechanical itself as a brain reduced to the work of calculation. Perhaps, as I have said, this is because plasticity is precisely the form of our world and because we are so immersed in it, so constituted by it, that we experience it without either thinking it or being conscious of it. We do this to such an extent that we no longer see that it structures our lives and sketches a certain portrait of power. We find here the poetical and aesthetic force that is the fundamental, organizing attribute of plasticity: its power to configure the world. Here again, Deleuze has perfectly analyzed this power by seeing in it the cinematographic function par excellence. The plasticity of the brain is the real image of the world. With a filmmaker like Alain Resnais, for example, "filmmaker of the brain," "landscapes are mental states" and mental states are universes and "cartographies,"[17] which renders them indiscernible and invisible as such. The films of Resnais, like those of Stanley Kubrick, display the identity of the brain and the world. We can think here of the noosphere of *Je t'aime, Je t'aime*, of the levels of structuration—which correspond to the forms of life of the different characters—of *Mon Oncle d'Amerique*, or of the giant computer in *2001: A Space Odyssey*.

The world configured in these films is not a centralized but a fragmented world, the faithful image of cerebral power, in which the dynamic "no longer works by totalization . . . but through continual relinked parcellings. . . . Hence the organic-cosmic bomb of *Providence* and the fragmentations through transformation of sheets in *Je t'aime, Je t'aime*. The hero is sent back to a minute of his past, but this is perpetually relinked into variable sequences."[18] The plasticity of time is inscribed in the brain. And we do not see it because it is a question of *our* time. We do not see it because it is a question of *our* world. We are perhaps always and necessarily blind, at first, to the political functioning and import of the brain-world (whence a certain reactivity, communally shared, to the films of Resnais). We are perhaps always and necessarily blind, at first, to our own cinema.

"The brain is adequate to the modern world," says Deleuze.[19] Perhaps precisely this adequation both blinds us to and explains and justifies the effects of the naturalization of the political and social by the neuronal, on the one hand, and the political and social effects of the descriptions of neuronal functioning, on the other. We recall that the most obvious transition point between the two domains is the crisis of centrality. But if we are living this crisis daily without really being able to think it, if we contrive to believe in a certain efficacy of the center (brain, machine . . .) that is perhaps because power—which hasn't been united for a long time, as Foucault endeavored to show us—has every interest in our imagining it that way. The screen that separates us from our brain is an ideological screen. By "screen" I mean both the cliché representations that I have just analyzed and the (only apparently) more "noble" resistances mounted against the neurosciences—more precisely against the cognitive sciences—by the majority of philosophers, psychoanalysts, and intellectuals in general. "Screen" also applies to the scientific descriptions themselves, which, pretending to lift the screen, really just reinforce it by producing no critical analysis of the worldview they implicitly drive.

Neuronal Man and the Spirit of Capitalism

Which worldview? Which world? The neo-liberal world, the world of global capitalism. The questioning of centrality, principal transition point between the neuronal and the political, is also the principle transition point between neuroscientific discourse and the discourse of management, between the functioning of the brain and the functioning of a company.

Revealing that the brain is neither a rigid structure nor a centralized machine is not enough to stave off the threat of alienation. In fact, neo-liberal ideology today itself rests on a redistribution of centers and a major relaxation of hierarchies. Domination and the crisis of centrality, in a merely seeming paradox, are perfectly matched with each other. The restructuring of capitalism (post-Fordist capitalism of the second industrial revolution) was accomplished at the price of substituting control by self-organization for planning decided and overseen by a formal centralized authority within the company. In the nineties, say Luc Boltanski and Eve Chiapello, "creativity, reactivity, and flexibility are the new watchwords," and "the bureaucratic prison explodes."[20] Or again, "the hierarchical principle is demolished and organizations become *flexible, innovative*, and highly *proficient*."[21] For this new organization, the *network* is the master term: current capitalism obeys the principle of mobile or "lean production" companies, "working as *networks* with a multitude of participants, organizing work in the form of teams or *projects*."[22] In such companies, one pays attention only to "the number, form, and orientation of connections."[23]

How could we not note a similarity of functioning between this economic organization and neuronal organization? How could we not interrogate the parallelism between the transformation of the spirit of capitalism (between the sixties and the nineties) and the modification, brought about in approximately the same period, of our view of cerebral structures? I have underlined the effect of the naturalization of the social attached to neuronal functioning. Boltanski and Chiapello confirm this: "This is how the forms of capitalist production accede to representation in each epoch, by mobilizing concepts and tools that were initially developed largely autonomously in the theoretical sphere or in the domain of basic scientific research. This is the case with neurology and computer science today. In the past, it was true of such notions as system, structure, technostructure, energy, entropy, evolution, dynamics, and exponential growth."[24] Like neuronal cohesion, contemporary corporate economic and social organization is not of a central or centralizing type but rests on a plurality of mobile and atomistic centers, deployed according to a connectionist model. In this sense, it appears that neuronal functioning has become the nature of the social even more than its naturalizing tool.

We must insist on this natural identity in returning to the notions of network, delocalization, and adaptability, and in observing how these operate in the two domains—cerebral and socioeconomic.

Networks

Cerebral organization presupposes the connection of neurons in networks, which are also called "populations" or "assemblies." In a network, there cannot be, by definition, a privileged vantage point. The network approach is necessarily local, never centralized or centralizing. Within the brain, writes Changeux, "the formal notion of a program finds itself replaced by an exhaustive description of properties, elements, geometry, and a communication network."[25] Thus, for example, the formation of what we call a "mental object"—an image or concept—requires a "*correlated, transitory* activity, both electrical and chemical, in a large population or 'assembly' of neurons in several specific cortical areas."[26] There is no longer a center but rather discrete assemblies of neurons forming mobile and momentary centers on each occasion. Organizational suppleness now goes hand in hand with authority and decision.

We know, moreover, that the zones of the brain serve many functions at once and can successively form part of many distinct functional networks. In other words, a given cerebral zone has no unique function: this is so for the "associative" areas of the brain. These multifunctional regions are activated in numerous cognitive tasks and form part of a different cerebral network each time. We find ourselves faced with a complex organization that no longer proceeds in a top-down fashion from transmission to reception to retransmission of information but functions according to different, extremely complex, interpenetrating levels of regulation. One therefore cannot attribute the directing function to just one of them: "The notion of the localization and cartographic organization of the brain must be modulated by the existence of a multitude of connections between brain regions as these have been identified by histology."[27] The phenomenon of the potentiation of circuits, discussed above, provides evidence that the nervous system is organized according to multiple interconnected functional spaces, always in movement and susceptible to self-modification.

It is obviously with reference to this type of functioning that today's management literature preaches work in "flexible, neural" teams,[28] and can claim that the manager "is not [or is no longer] a (hierarchical) boss, but an integrator, a facilitator, an *inspiration*, a unifier of energies, an *enhancer of life*, meaning, and autonomy."[29] The team has faith in him "inasmuch as he proves to be a *connector*, a *vector*, who does not keep the information or contacts gleaned from the network to himself but redistributes them among team members. 'Tomorrow's manager should make sure that information is shared, that it irrigates the firm thoroughly.'"[30] If it is true that the "boss" has always been compared to the "brain," we can see clearly that the neuronal manager no longer has the same style of government or command as the cerebral C.E.O. Ideally, the boss can refrain—at least in appearance—from giving orders: in principle, "the leader has no need to command," because the personnel are "self-organized" and "self-controlling."[31] He transmits, distributes, and modifies connections by potentiating or depressing them according to circumstances and needs, without being identifiable with or assigned to a fixed post. Thus, "the manager is clearly the network man. His principal quality is his mobility, his ability to displace himself."[32] The abolition of centrality goes hand in hand with the capacity to delocalize oneself.

Delocalization

We have just seen that connections between different regions of the brain allow us to think a certain delocalization of cerebral activities. In effect, it seems that the localizations described by anatomists and neurologists are no longer what they were: they no longer form a rigid topography but are included in networks made and unmade as a function of the cognitive task in which the subject is engaged.[33] New neuroimaging methods allow us to visualize the zones of the brain involved in the realization of cognitive tasks. Yet the ensemble of zones involved in this type of task (the classic cerebral localizations) takes the form, as we have seen, of a temporarily activated network, somehow recruited by the task to be accomplished and the cognitive context in which it is accomplished. The realization of another task would give the network a different configuration, in which some of the preceding localizations would find themselves grouped differently. The same region can contribute to the realization of different functions.[34] The mental object, in turn, says Changeux, has an organization "both local and delocalized."[35] The primary qualities of assemblies of neurons are their mobility and their multifunctionality.

But aren't these qualities also those expected today of the individual in the working world? Shouldn't we become polyvalent, accepting the law of delocalization by making ourselves available, showing ourselves to be without attachment, ready to break old ties, to create new ones? In a company, write Boltanski and Chiapello, "valued staff members are those who succeed in working with very different people, prove themselves open and flexible when it comes to switching projects, and always manage to adapt to new circumstances."[36] Today the emphasis is clearly put on polyvalence more than on craft, on the multiplication of encounters and potentially reactivizable temporary connections, on belonging to diverse groups. Capitalism obviously—implicitly *and* explicitly—refers to neuronal functioning as it pretends "to replace essentialist ontologies with open spaces without borders, centers, or fixed points, where entities are constituted by the relations they enter into and alter in line with the flows, transfers, exchanges, permutations, and displacements that are the relevant events in this space."[37] This happens to such a degree that anchorings in a space or a region, attachment to family or a domain of specialization, and overly rigid fidelity to self appear incompatible with what today is called "employability." One must always be *leaving* in order to survive, that is to say, in a certain sense, in order to *remain*.[38]

Adaptability

Whoever says "employability" clearly says *adaptability*. "Employability" is a neo-management concept that indicates "the capacity to respond to a world in motion" by a supple use of abilities, which supposes we do not focus on one and only one skill, just as a cortical region does not participate in one and only one function. "Far from being attached to an occupation or clinging to a qualification, the great man proves *adaptable* and *flexible*, able to switch over from one situation to a very different one and adjust to it; and *versatile*, capable of changing activity or tools, depending on the nature of the relationship entered into with others or with objects."[39] It is a question of not locking oneself into a specialization while still having a specific skill to offer.

"Employability" is synonymous with *flexibility*. We recall that flexibility, a management watchword since the seventies, means above all the possibility of instantly adapting productive apparatus and labor to the evolution of demand. It thus becomes, in a single stroke, a necessary quality of both managers and employees. If I insist on how close certain managerial discourses are to neuroscientific discourses, this is because it seems to me that the phenomenon called "brain plasticity" is in reality more often described in terms of an economy of flexibility. Indeed, the process of potentiation, which is the very basis of plasticity, is often presented simply as the possibility of increasing or decreasing performance. Very often, the brain is analyzed as personal capital, constituted by a sum of abilities that each must "invest optimally," like an "ability to treat one's own person in the manner of a text that can be translated into different languages."[40] Suppleness, the ability to bend, and docility thus appear to join together in constituting a new structural norm that functions immediately to exclude.

Social "Disaffiliation" and Nervous Depression: The New Forms of Exclusion

In effect, anyone who is not flexible deserves to disappear. In *The Fatigue of Being Oneself: Depression and Society*, a work dedicated to nervous depression and the new psychiatry, sociologist Alain Ehrenberg shows that the frontier separating psychical suffering and social suffering is thin. Depression is merely a form of what another sociologist, Robert Castel, calls "disaffiliation." In both cases, it is a question of suffering from exclusion, articulated as so many illnesses of flexibility. The depressed person, like the "social failure," evidently suffers from a lack of "employability" and adaptability. The coincidence between current psychiatric discourse, characterized by a clear tendency toward the "biologization" of psychical or mental disturbance, and the political discourse of exclusion, which presents the disaffiliated as individuals "with broken connections," is striking. Before coming to the necessary distinction we must work out between a simply flexible identity and a truly plastic identity—a distinction resting on a theory of transformation—we should pause a moment on the question of this suffering. About this topic psychiatrists, neurobiologists, and politicians all advise the same thing: modify the neuronal (the "network") to differently configure oneself; amplify connections to reinstate mental and behavioral "plasticity."

"Structural and functional brain imaging," we read in a medical brochure,

> have . . . shown that depressive episodes are accompanied by anatomo-functional correlates in certain brain regions, more precisely, at the level of networks involving the prefrontal cortex, hippocampus, and amygdala. On this basis, it has been possible to identify signs of hippocampic atrophy associated with hyperactivity of the corticotropic axis in recurrent depressions, as well as in post-traumatic stress disorder. These claims have led to the hypothesis of the

neurotoxicity of anxious and depressive episodes. Moreover, the neurological study of cerebral structures has revealed signs of neuronal, axonal, and dendritic atrophy, with diminution of synaptic connections and of nervous tissue.[41]

Thus depression, indeed, psychical suffering in general is associated with a diminution of neuronal connections (as if the concept of long-term depression had a literal sense). This diminution usually corresponds to an inhibition.

The depressed person is indeed frequently "apathetic," characterized by "holding back, stiffening, braking, and suspension of activity."[42] Nevertheless, "mental disturbance no longer concerns a person's difficulties; [it becomes] an illness that cuts a patient off from his aspect as agent."[43] This redefinition of an ill person as cut off from his possible actions on the cognitive level as well as on the emotional and purposive level corresponds to the biologization or "re-biologization" of disturbance mentioned above. From such a perspective, therapy consists first and foremost in analyzing the mechanisms blocking transmission of information in the neuronal systems. Antidepressants, in their great diversity, all seek to stimulate neurochemical transmission, with the avowed goal of "restoring and protecting the plastic capacities of the brain."[44] But plasticity ought not to be confused, as we will see, with the mere capacity to act.

Once again, it is not a matter of criticizing psychiatric reductionism in the name of a supposed "freedom" of psychism. To deny the neurological foundation of depression, to deny the therapeutic power of certain molecules, would be absurd and vain. Neuropsychiatry is without question one of the most promising disciplines today, and I avidly follow the molecular adventure of psychopharmaceuticals. It is therefore not a question of pitting the nobility of "classical" psychoanalysis against the baseness of psychiatry, but of seeing how a certain conception of flexibility—paradoxically driven by the scientific analysis of neuronal plasticity—models suffering and allows the identification of psychical illness and social illness.

Today these two types of disturbance tend to be conflated. One must see clearly that, for all intents and purposes, "the workplace is the antechamber of nervous depression."[45] The absence of centrality and hierarchy evoked above, the absence of clear and localized conflict, and the necessity of being mobile and adaptable constitute new factors of anxiety, new psychosomatic symptoms, new causes of severe neurasthenia. "In business," explains Alain Ehrenberg:

> the (Taylorian or Fordist) disciplinary models of human resources management are on the decline, in favor of norms that encourage autonomous behavior, even for personnel at the bottom of the hierarchy. . . . Modes of regulation and domination of the workforce are now based less on mechanical obedience than on initiative: responsibility, the capacity to evolve, to form projects, motivations, flexibility, etc. . . . The model imposed on the worker is no longer that of the man-machine of repetitive labor, but that of the entrepreneur of flexible labor.[46]

Thus a depressive is a sick person who cannot stand this conception of a "careerist" whose very existence is conceived as a business or a series of projects.

Ehrenberg continues:

> Whatever domain one considers (company, school, family), the rules of the world have changed. They are no longer obedience, discipline, and conformity to morals, but flexibility, change, reaction time, etc. The

demand for self-mastery, affective and psychical suppleness, and capacities for action force each person to adapt continuously to a world without continuity, to an unstable, provisional world in flux, and to careers with ups and downs. The legibility of the social and political game is muddied. These institutional transformations give the impression that everyone, including the most fragile, must take up the task of *choosing everything* and *deciding everything.*[47]

Such a situation surely creates a certain vulnerability, a new precariousness, a new fragility. The difficulty in experiencing a conflict voids the psyche and in effect replaces neurosis with "the fatigue of being oneself."

Robert Castel thematizes

the presence, apparently more and more insistent, of individuals who virtually drift about within the social structure, and who populate interstices of society without finding any established position within it. Vague silhouettes, at the margins of labor and at the frontiers of socially consecrated forms of exchange— the long-term unemployed, inhabitants of abandoned suburbs, recipients of a national minimum income, victims of industrial downsizing, young people in search of employment who carry themselves from place to place, from menial jobs to temporary work— who are these people, where did they come from, and what will become of them?[48]

This vocabulary of drifting, of lack of place, of wandering, obviously recalls that of depression, inhibition, or anxiety. The phrase "social question" in the title of Castel's book *From Manual Workers to Wage Laborers: The Transformation of the Social Question* means "a concern about a society's ability to maintain its own cohesion."[49] Yet how could we not think that there is conjointly a psychiatric question, bearing the same sense, which testifies (and perhaps only testifies) to a worry about social cohesion? How could we not think that depressive or disaffiliated individuals represent threats of turbulence, of breaks in transmission in the fluidity of the network? "In a connectionist world, where high status presupposes displacement," write Boltanski and Chiapello, "the big shots derive part of their strength from the immobility of the little people, which is the source of their poverty. . . . Everyone thus lives in a state of permanent anxiety about being disconnected, rejected, abandoned on the spot by those who displace themselves."[50] This anxiety creates precariousness, that is to say, "the increasingly drastic privation of links and the progressive emergence of an inability not only to create new links but even to maintain existing links (separation from friends, breaking of family ties, divorce, political abstentionism)."[51] This lack of ties and this risk of being cut off appear as threats that one must contain or ward off at any cost to maintain the cohesion of the community.

Hence to heal means to reintegrate, to restore flexibility. When it first appeared, Prozac was presented as a "mood raiser" and an "action facilitator." In his book *Listening to Prozac*, Peter Kramer develops a critical reflection on the type of "self" that "today's high-tech capitalism" endorses as its condition of possibility: "Confidence, flexibility, quickness, and energy . . . are at a premium."[52] Prozac allows one to obtain these goods at a low cost, not only because this medication is not expensive, but also because it allows one to avoid the psychical cost of acquiring these values. Mood medications, or "thymoregulators," thus seem to have the function of reducing vulnerability, chronic disturbance, and psychical precariousness by targeting the neuronal networks involved in initiative, stimulation, dynamism, and well-being. Medications should give back the appetite for mobility, the capacity to rid oneself of rigidity and of fixity in one's identity.

Thus it is no longer possible to distinguish rigorously on an ideological level between "popularly" accessible neuroscientific studies and the literature of management—including medical management. Think, for example, of how Alzheimer's patients are described. An Alzheimer's patient is the nemesis of connectionist society, the countermodel of flexibility. He is presented as a disaffiliated person: errant, without memory, asocial, without recourse. One observes in his brain a thinning of connections, the accumulation of fibrils inside neurons, and the presence of senile plaques—all factors contributing to rigidification and loss of suppleness, which, paradoxically, lead to a chaotic wandering.[53] In how rigidity prevents initiative here, one can see an obvious relation between, on the one hand, the image constructed and conveyed of such a patient and, on the other, those constructed of the homeless, illegal immigrants, or unemployed persons about to be kicked off the dole. In fact, it is no longer possible to distinguish rigorously on an ideological level between those suffering a neurodegenerative disorder and those with major social handicaps.

As we have observed, any vision of the brain is necessarily political. It is not the identity of cerebral organization and socioeconomic organization that poses a problem, but rather the unconsciousness of this identity. The persistent use of long-defunct technological models to represent the brain bars access to a true understanding of cerebral function and justifies our lack of interest in it. The representations/obstacles of a rigid encephalon, cut off from thought, cut off from the essential, are precisely what induce us to keep the brain away from itself, to separate it from what it is: that is, the essential thing, the biological, sensible, and critical locus of our time, through which pass, one way or another, the political evolutions and revolutions that began in the eighties and opened the twenty-first century. At bottom, neuronal man has not known how to speak of himself. It is time to free his speech.

Indeed, without this freeing, neuroscientific discourse will have the sole consequence—beyond medical advances—of unwittingly producing criteria, models, and categories for regulating social functioning and increasing daily the legitimation of the demand for flexibility as a global norm. To produce consciousness of the brain is not to interrupt the identity of brain and world and their mutual speculative relation; it is just the opposite, to emphasize them and to place scientific discovery at the service of an emancipatory political understanding.

On the one hand, neuronal functioning as it is described today quite closely resembles a democracy: mutual support (reparation), freedom of choice (one somehow constructs one's own brain), a crossing point between the public and the private (the interaction of the outside and the inside), belonging to many spheres, mobility, openness, availability, autonomy, absence of hierarchy between the network elements, and equality of function. (By contrast, the models of the central telephone exchange and the computer continue to evoke the old Soviet system or *Brave New World*.) In one sense, progress in the neurosciences has made possible the political emancipation of the brain. On the other hand, the scientific description of brain plasticity produces, while taking its inspiration from, an extremely normalizing vision of democracy, in that it accords an overly central role to the absence of center, a too rigid prominence to flexibility, that is to say, to docility and obedience. Producing a consciousness of the brain thus comes down to producing the conditions of possibility for a new world of questioning: Can the description of brain plasticity escape the insidious command of the New World Order? Can it introduce something like a resistance within this very order? *Can plastic brains measure the limits of their flexibility?*

3

"You Are Your Synapses"

My approach to these questions may, at first glance, seem surprising. I have just brought out the most visible points of transition between the neuronal and the political—in other words, between the biological and the social. I have shown that the concept of flexibility, the transitional or transborder concept par excellence, also concealed this transition's theoretical conditions of possibility. In closing, therefore, let us linger over this concept. But how? Here comes the real surprise: we will now turn to what constitutes the chief affirmation of the neurosciences in general, and of the cognitive sciences in particular—the certainty that there exists a perfect continuity between the neuronal and the mental.

The current state of research and observation allows cognitive scientists to conclude that thought, knowledge, desires, and affects all proceed on a neuronal, that is to say, biological, basis, and that the mental images constituting the life of the mind are indeed formed in the brain. This chief affirmation, which is the basis for all "reductions" (in other words, the basis for assimilating the mind to a natural datum), is at once the strongest and the weakest point of neuroscientific discourse in general. It is the strongest because, even if sometimes shocking, it is incontestably, whatever we think of it, the expression of a real advance: it has enabled us to approach phenomena such as memory, perception, learning—even psychical and behavioral problems—more and more precisely and objectively. In the most general way, it constitutes a new approach to the subject by affirming the existence of a "neuronal self." It is the weakest because the certainty of the continuity between the neuronal and the mental can obviously never be a strictly scientific postulate. It necessarily constitutes a philosophical or epistemological position and such positions are not always clearly articulated. I will therefore attempt to question the presuppositions attached to this continuity, not to contest it in itself but to show that its development and function are precisely discontinuous—that it is, in other words, a question of a complex continuity.

Logically, how are we to explain what relation could exist between such a study and the political, social, and economic questions raised above? The answer is the following: interrogating the transition from the neuronal to the mental leads us to interrogate the very core of cerebral functioning, the transition from the biological to the cultural, from the strictly natural base of the mind to its historical—and thus also, necessarily, its political and social—dimension. Reinvestigating the question of the transition from the neuronal to the political within the field of the neuronal itself should allow us, through a strategy of redoubling, to bring out the theoretical mediations, transitions, indeed, the theoretical holes likely to unsettle the very concept of continuity, and in so doing to perturb flexibility. We shall thereby be able to grasp the distinction between what is truly liberating in this new definition of self and what within it remains a subjugating power. This "weighing in the balance" will require a critical confrontation between flexibility and plasticity.

The "Synaptic Self" or "Proto-Self"

We will begin with the concept of the subject or "self" supported by contemporary neuroscientific discourse. We will intentionally not stop saying "we," even when simply asking the question "What should we do with our brain?" For who is this "we," and what relation does this "we," the very possibility of saying "we," have to the brain?

For most neurobiologists today, the brain is not a simple "organ" but the very possibility of linking, the fundamental organic coherence of our personality, our "we," a consideration that tends to blur the line between the nervous system and the psyche. Prominent neurobiologists such as Antonio Damasio and Joseph LeDoux now clearly affirm this point: consciousness is nothing other than "how the owner of the movie-in-the-brain emerges within the movie,"[1] and, as a result, we need to grasp "the essence of a person in the brain."[2] To examine this essence, we will follow the demonstrative order adopted by LeDoux in his book *Synaptic Self.* "In previous chapters," he writes, "we've seen how neuronal circuits are assembled during development, and how these circuits are modified when we learn and remember. Now we will begin to use this basic information about circuits and their plastic properties to explore broader aspects of mental function, that is, to begin to develop a neurobiological view of the self."[3]

So what in fact is this synaptic "self," or "proto-self," as Damasio chooses to call it? Why does the analysis of brain plasticity necessarily drive us to posit its existence? To what extent is it possible to determine a personal identity on the basis of neuronal configurations and so to consider that the brain is the first and most fundamental form of subjectivity? The response to all these questions seems elementary: "My notion of personality," says LeDoux, "is pretty simple: your 'self,' the essence of who you are, reflects patterns of interconnectivity between neurons in your brain. . . . Given the importance of synaptic transmission to brain function, it should practically be a truism to say that the self is synaptic."[4] Or again: "The essence of who you are is stored as synaptic interactions in and between the various subsystems of your brain. As we learn more about the synaptic mechanisms of memory, we learn more about the neural basis of the self."[5]

Thus an awareness of synaptic plasticity leads scientists to advance the thesis of a neuronal personality. The "self" is a synthesis of all the plastic processes at work in the brain; this permits us to hold together and unify the cartography of networks already mentioned. "The fact that plasticity does occur in so many brain systems," we read in *Synaptic Self*, "raises . . . interesting questions. How does a person with a coherent personality—a fairly stable set of thoughts, emotions, and motivations—ever emerge? Why don't the systems learn different things and pull our thoughts, emotions, and motivations in different directions? What makes them work together, rather than as an unruly mob?"[6] It is the "self," incontestably, that allows for this grouping and linking.

The "proto-self," or "primordial self," explains Damasio, covers "the ensemble of brain devices which continuously and *nonconsciously* maintain the body within the narrow range and relative stability required for survival. These devices continually represent, *nonconsciously*, the state of the living body, along its many dimensions."[7] The proto-self is thus primarily a form of *organic representation of the organism itself* that maintains its coherence: "as far as the brain is concerned, the organism . . . is represented by the proto-self. The key aspects of the organism . . . are . . . provided in the proto-self: the state of the internal milieu, viscera, vestibular system, and musculoskeletal frame."[8] This base that represents itself to itself is the very condition of life. Without it there is no possible survival and no consciousness. Indeed, the nonconscious processes at work in the proto-self are the very conditions of consciousness: "the proto-self is the nonconscious forerunner for the levels of self which appear in our minds as the conscious protagonists of consciousness: core self and autobiographical self."[9] The proto-self is thus a "preconscious biological precedent" out of which alone can be developed the sense of self (core self, "core consciousness," or "I") and the temporal and historical permanence of the subject (autobiographical self, "invariant aspects of an individual's biography").[10]

One can see that the notion of biological precedence leads directly to that of the continuity between the neuronal and the mental. Indeed, core consciousness and autobiographical consciousness are formed from, and emerge from, the proto-self in a progressive manner, without rupture or leap. How is this continuity possible? Here is the most interesting and most subtle point of the analysis: through modification of the primitive or primordial representational function that is the work of the proto-self. Indeed, one must suppose that the "proto-self" presents itself as "a coherent collection of neural patterns which represent the state of the organism, moment by moment, at multiple levels of the brain."[11] Thus there actually is, contrary to Bergson's claim, a self-representation of the brain, an auto-representation of cerebral structure that coincides with the auto-representation of the organism. This internal power of representation inherent in neuronal activity constitutes the prototypical form of symbolic activity. Everything happens as if the very connectivity of the connections—their structure of reference, in other words, their semiotic nature in general—represents itself, "maps" itself, and precisely this representational activity permits a blurring of the borders between brain and psyche.[12]

The brain thus informs itself about its own state to the extent that it is informed about the state of the organism, an economy of transmission assured by a play of "signals" that Damasio calls "impulses." This elementary conversation, which constitutes one of the primary activities of the nervous system, is still called "nonconscious."[13] The ongoing modification of this first cerebral habitus gives rise to more and more complex, and more and more stable "maps." The construction of the link to the object demands the formation of images, or "second-order maps," and thereafter of signs. In detail, the stages are the following: "the nonconscious neural signaling of an individual organism begets the *proto-self*, which permits *core self* and *core consciousness*, which allow for an *autobiographical self*, which permits *extended consciousness*. At the end of the chain, *extended consciousness* permits *conscience*."[14] From one end of the chain to the other, Damasio explains, one must assume that the brain somehow recounts its own becoming, that it elaborates it in the form of an "account."

Within the cerebral structure there is something like a poetic activity or a wordless recitative function:

> The account describes the relationship between the changing proto-self and the sensorimotor maps of the object that causes these changes. In short: As the brain forms images of an object—such as a face, a melody, a toothache, the memory of an event—and as the images of the object *affect* the state of the organism, yet

another level of brain structure creates a swift and nonverbal account of the events that are taking place in the varied brain regions activated as a consequence of the object-organism interaction. The mapping of the object-related consequences occurs in first order neural maps representing proto-self and object; the account of the *causal relationship* between object and organism can only be captured in second-order neural maps. Looking back, with the license of metaphor, one might say that the swift, second-order nonverbal account narrates a story: *that of the organism caught in the act of representing its own changing state as it goes about representing something else.*[15]

From the "proto-self" to "conscience" there thus develops an extensive "re-representation *of the nonconscious proto-self in the process of being modified.*"[16] This process corresponds to the translation of neuronal patterns into mental patterns. The latter (thus "images" and "signs") constitute the elementary life of the three domains of cognition, emotion, and motivation, the fundamental tripartite division of the mind. Damasio affirms that "the brain makes neural patterns in its nerve-cell circuitry and manages to turn these neural patterns into the explicit mental patterns which constitute the highest level of biological phenomenon,"[17] and which he likes to call, succinctly, images—visual images, auditory images, tactile images, and so forth, images that can convey any object, any relation, concrete or abstract, any word, and any sign.

The transition from the neuronal to the mental is confirmed by the fact that it is impossible to distinguish the two domains rigorously and absolutely. If, in effect, there is a kind of subterranean representational activity in the brain, this already signifies that neurons, through "being in connection," are already available for, and already disposed toward, meaning. In the same way, meaning, or symbolic activity in general, depends strictly on neuronal connectivity.

"Lost in Translation:" From the Neuronal to the Mental

Fascinating as they may be, these analyses remain insufficient on many points. Despite the apparent assurance and certitude that govern the discourse of the "adherence" of the mental to the neuronal, the process of the "translation" of the givens from one domain to the other remains obscure. No matter what is said about it, this "translation," for all its plausibility with respect to its function, remains questionable with respect to its lawfulness: it has never managed to be truly constituted as a law, nor to acquire thereby the value of a universal. No one today is in a position to prove that all cognitive, emotional, or practical activities are the reformulated and resystematized equivalents of neuronal configurations. As LeDoux notes, "I'll state unashamedly from the start that we can't, at this point, go all the way in formulating a complete synaptic theory of personality."[18]

If there is always a mental dimension to the neuronal and a neuronal dimension to the mental, then we must suppose that this continuity is in some way itself at once neuronal and mental, biological and cultural, or, if we anticipate the "translations," at once an object of observation and an interpretive postulate. The continuity from the neuronal to the mental, let us recall, is in essence a theoretical mixture, at once experimental and hermeneutic, as Damasio's recourse to the metaphors of narrative and text reveals. Thus, the space and the cut that separate the neuronal from the mental, or the proto-self from different forms of consciousness, are comparable not to synaptic gaps, openings that permit passage without ever hindering it, but rather to theoretical fissures that, in order to be minimized, require that scientific explanation be relayed by *interpretation*.

In saying this, I in no way presume to contest the hypothesis of neuronal and mental continuity or to play the game of antireductionism. It seems important simply to insist that, when this theoretical fissure is not recognized as such—as in the great majority of neuroscientific discourses—it runs the risk of being overwhelmed by brute, naïve ideology.

It is of course entirely possible to postulate that the organism, "as a unit, is mapped in the organism's brain, within structures that regulate the organism's life and signal its internal states continuously . . . [and that] all of these neural patterns can become images."[19] The problem remains that of grasping the nature of this becoming, which permits the transformation of the proto-self into a conscious element. Certainly, as we have just seen, Damasio proposes an explication and a metaphorics of this transformation. The idea of a nonconscious process of metabolic representation is extremely interesting: effectively, it lets us formulate the hypothesis of a metamorphic fluidity assuring the synthesis of the cerebral and the psychical. But the entire question lies in the modality of this "synthesis," the conditions of possibility of this fluidity.

What, finally, is the ultimate source of this metabolism or this cerebral/mental converter? No response advanced by the neurologists is truly satisfying. Basically, "change," "translation," "account," and "narrative" are too vague and, without further analysis, do not let us grasp simultaneously the transition from one level of organization to another (from the neuronal to the mental), the transition from one organizational regime to another (from the self-conservation of the proto-self to the exploratory activity of consciousness), and the transition from one organizational given to another (the proto-self is a genetic given; the self that manipulates images and signs is a biological-cultural given). We do not truly know what originally makes these transitions possible: Are they biologically programmed? Are they the fruit of experience or of individual history? Are they the result of both?

In making more precise his definition of the term *non-conscious*, Damasio declares:

In fact, the list of the 'not-known' is astounding. Consider what it includes: (1) all the fully formed images to which we do not attend; (2) all the neural patterns that never become images; (3) all the dispositions that were acquired through experience, lie dormant, and may never become an explicit neural pattern; (4) all the quiet remodeling of such dispositions and all their quiet re-networking—that may never become explicitly known; and (5) all the hidden wisdom and know-how that nature embodied in innate, homeostatic dispositions.[20]

It is thus legitimate to ask why certain neuronal patterns never become images, why certain dispositions never become schemas. What remains mysterious (and we cannot be satisfied here by evoking "the wisdom of nature") is therefore the deep structure of transformation, the transition from a universal self, not yet particularized, to the singular self, to that which I am, that which we are.

Not to interpret is still to interpret. By wishing not to construct a hermeneutic schema capable of explaining, at least provisionally, the relations between the neuronal and the mental, by wishing not to recognize the necessarily meta-neurobiological dimension of that schema, one exposes oneself, whether one recognizes it or not, to ideological drift—for example, and above all, to that of mental Darwinism or psychological Darwinism.

At this point we encounter yet again the political, economic, and social questions developed above. According to the "logic" of these "Darwinian" positions, only those neuronal configurations capable of survival, thus those capable of being the "best," the highest "performing," would be converted into images. Only the most "useful" synaptic connections would be modulated or reinforced.[21] There would be at the very heart of the self a selection oriented toward efficacy. Damasio affirms that "our attitudes and our choices are, in no small part, the consequence of the 'occasions of personhood' that organisms concoct on the fly at each instant."[22] But it would seem that certain persons have more "occasions" than others, since Damasio himself speaks of qualitative differences in individuality, referring to the "personalities that appear to us as most harmonious and mature,"[23] on the basis of the number and neuronal richness of the connections that underlie them.

But what could be meant by this supplement of maturity and harmony characteristic of certain "selves," if not an excess of power or capacity for success, a higher number of chances to occupy a dominant position? And to what or to whom, inversely, could "nonharmonious" or "immature" personalities correspond, if not, in one way or another, to the disaffiliated we evoked above? Where is the dividing line between the two? This forces us back to the problem of the "transition." If we are from the start a nonconscious proto-self always "in a process of being modified," how is this modification effected? Does it proceed solely by natural selection (or cultural selection, which amounts to the same thing)?[24] Must we assume an original flexibility that, by adaptive selection, forms personality? Must we postulate the suppleness of a primordial self that can (or even ought to) bend to the working of the simultaneously biological and cultural barrage to which it is subjected?

These questions are fundamental. Awakening a consciousness of the brain, as we are trying to do, means awakening a consciousness of the self, a consciousness of consciousness, if you will, which is also to say a comprehension of the transition from the neuronal to the mental, a comprehension of cerebral change. The brain is our work and we do not know it. The brain is constituted by modifications of modifications, of "re-representations," and we do not know it. The brain owes its vitality to a perpetual change in plasticity (which is also to say a plasticity of change itself) and we do not know it. In setting these points aside in order to discuss only the results, neurobiologists and cognitive scientists contribute to confirming the diffuse and highly paradoxical feeling that the brain is the locus of an absence of change and that we cannot in reality do anything about it, do anything with it, other than letting selection have its way. But really, what's the point of having an all-new brain if we don't have an all-new identity, if synaptic change changes nothing? And what do we get from all these discourses, from all these descriptions of neuronal man, from all these scientific revolutions, if not the absence of revolution in our lives, the absence of revolution in our selves? What new horizons do the new brains, the new theoreticians of the brain, open up?

About Antonioni's cinematic work, Deleuze declares: "Antonioni does not criticize the modern world, in whose possibility he profoundly 'believes': he criticizes the coexistence in the world of a modern brain and an exhausted body."[25] We could say in the same way that today we live the "coexistence of a modern brain and an exhausted identity." All the fascinating discoveries of the neurosciences remain a dead letter for us, never managing to destroy our old representations of the brain (for example, that of the machine brain), because they are incapable of unleashing possibilities, of unleashing new ways of living and—why be afraid of the word?—new ways to be happy. It must be acknowledged that neuronal liberation has not liberated us. Long-term potentiation and depression cannot be the first and last words on the plasticity of a self, in other words, on its modification by experience.

Even if it is fascinating to observe aplysias, we cannot spend our time in ecstasies over slugs. Nor in asking ourselves, as certain popularizing scientific magazines often invite us to do: "How does the brain activity of a mathematician differ from that of an architect?" "What cerebral regions are active when a lawyer is preparing his arguments?" "Can we teach people to activate the appropriate cerebral regions to improve their performance?" or even "Will it soon be possible to read thoughts?"[26] All of this is, at bottom, a matter of perfect indifference to us, and our self itself—as much as our body—is exhausted by such an absence of perspective. A sad story for a sad subject, never granted an understanding of its own transformation.

We must acknowledge an enormous discrepancy between the descriptive and the prescriptive scope of neuroscientific discourses. We must acknowledge an enormous discrepancy between, on the one hand, all the promises for the future and desires for another history and another life, desires aroused by this wholly new vision of the brain, by this continent known as cerebral plasticity, and, on the other hand, the tiny political, philosophical, and cultural space in which these promises can at once be theoretically deployed and realized. Once again, it seems that the neuronal revolution has revolutionized nothing *for us*, if it is true that our new brains serve only to displace ourselves better, work better, feel better, or obey better. The synthesis of the neuronal and the psychical thus fails to live up to its task: we are neither freer, nor smarter, nor happier. "The individual today," says Ehrenberg, "is neither sick nor healed. He is enrolled in multiple maintenance programs."[27] Do we want to continue to be "chronically healthy" in this way?

How can we fail to see that the only real view of progress opened by the neurosciences is that of an improvement in the "quality of life" through a better treatment of illness?[28] But we don't want these half-measures, what Nietzsche would rightly call a logic of sickness, despairing, and suffering. What we are lacking is *life*, which is to say: *resistance*. Resistance is what we want. Resistance to flexibility, to this ideological norm advanced consciously or otherwise by a reductionist discourse that models and naturalizes the neuronal process in order to legitimate a certain social and political functioning.

Another Plasticity

If we can accept the idea that personality derives from a constellation of established connections, then we can also accept that personality is reformable or re-formable. If this is so, are this reformation and this re-forming without limits, or do they have some capacity to resist an excess of polymorphism? Here we rejoin the point of confrontation between flexibility and plasticity.

In order to answer all the questions that I have tried to raise from the outset, it strikes me as absolutely necessary to introduce into the register of cerebral plasticity discussed in the first chapter—in addition to developmental plasticity, modulational plasticity, and reparative plasticity—a fourth type of plasticity, never as yet envisaged by neuroscientists, that would enable and qualify the formation of the singular person on the basis of the neuronal matrix. An intermediate plasticity of some kind, situated between the plasticity of the "proto-self" and that of the conscious self.[29] Once again, I in no way seek to contradict the thesis of the transition from the neuronal to the mental or to affirm the existence of an assumed incommensurability between one domain and the other. I do not adopt this "antireductionist" position but rather think that a reasonable materialism should accept the necessary mediation of the idealization of self—that the position of neuronal materialism, which I adopt absolutely, should elaborate a central idea, or theory, of the transition. But this plasticity of transition, omitted from neurobiology treatises, this plasticity connecting protoplasticity to experiential plasticity, should constitute this theoretical bedrock, this idea or this idealization. "You are your synapses": I have nothing against this sentence. I simply want to understand the meaning of "being" here.

For that, we must arrive at an intermediate plasticity, a plasticity-link that is never thought of or recognized as such, allowing us to elaborate a true dialectic of the auto-constitution of the self. This is what we must discern, as did Freud in his day by analyzing the type of transformation enabling the transition from the neuronal to the psychical, the latter never being, in a certain sense, anything more than the metamorphosis of the former.[30] If we do not think through this transformation or this plasticity, we dodge the most important question, which is that of freedom. If, in effect, the life of the brain is played out between program and deprogramming, between determinism and the possibility of changing difference, then the transition from the proto-self to the self is indeed the transition from the undifferentiated to the possibility of a transdifferentiation of self—the self, between receiving and giving form, being at once what one inherits and what one has created. But we cannot settle for a neutral description of the three types of plasticity discussed in the first chapter; we must also propose a model of their interaction and the joint dynamics of their genesis: how modulation links up with modeling, how reparation changes its meaning with experience, and how these interactions construct a free personality or singularity. But in order to understand such a construction, we must leave the domain of pure description and agree to elaborate a theoretical petition, once again necessarily meta-neurobiological, as Freud wrote, feeling the need to go behind or beyond, a metapsychology.

The Upsurge and Annihilation of Form

Plasticity is situated between two extremes: on one side, the taking on of form (sculpture, molding, fashioning of plastic material); on the other, the annihilation of form (plastique, detonation). Plasticity deploys its meaning between sculptural modeling and deflagration—in other words, explosion. Let us now examine this last meaning. Essentially, today we must think this double movement, contradictory and nonetheless indissociable, of the emergence and disappearance of form. At the core of the constant circulation between the neuronal, the economic, the social, and the political that characterizes Western culture today, the individual ought to occupy the midpoint between the taking on of form and the annihilation of form—between the possibility of occupying a territory and accepting the rules of deterritorialization, between the configuration of a network and its ephemeral, effaceable character. We live in an epoch in which identity is defined no longer as a permanent essence but as a process of autoconstitution or "fashioning," to reclaim the term used by Foucault, a process at whose heart a multiplicity of possible figurations unfolds. Today everyone lives multiple lives, at the same time and successively.

Self-fashioning implies at once the elaboration of a form, a face, a figure, and the effacement of another form, another face, another figure, which precede them or are contemporaneous with them. On the one hand, the coincidence between formation and disappearance of form is diachronic: a past form cedes place to a new form, and one thus changes identity or "self" in the course of time. On the other hand, the coincidence between formation and disappearance of form is synchronic: the threat of the explosion of form structurally inhabits every form. All current identity maintains itself only at the cost of a struggle against its autodestruction: it is in this sense that identity is dialectical in nature.

What does this mean? The plasticity of the self, which supposes that it simultaneously receives and gives itself its own form, implies a necessary split and the search for an equilibrium between the preservation of constancy (or, basically, the autobiographical self) and the exposure of this constancy to accidents, to the outside, to otherness in general (identity, in order to endure, ought paradoxically to alter itself or accidentalize itself). What results is a tension born of the resistance that constancy and creation mutually oppose to each other. It is thus that every form carries within itself its own contradiction. And precisely this resistance makes transformation possible.

The auto-constitution of self obviously cannot be conceived as a simple adaptation to a form, to a mold, or to the received schemata of a culture. One is formed only by virtue of a resistance to form itself; polymorphism, open to all forms, capable of donning all masks, adopting all postures, all attitudes, engenders the undoing of identity. Rather than displaying a real tension between maintenance and evolution, flexibility confounds them within a pure and simple logic of imitation and performance. It is not creative but reproductive and normative.

Life and Explosion: Homeostasis and Self-Generation

Let us return to the problem of the transition from the neuronal to the mental. The dialectical nature of identity is rooted in the very *nature* of identity, that is to say, in its biological foundation. Indeed, in adopting the thesis of a neuronal self, I would postulate that it, too—indeed, it above all—is structured by the dialectical play of the emergence and annihilation of form, that the historico-cultural fashioning of the self is possible only by virtue of this primary and natural economy of contradiction.

The transition from the neuronal to the mental supposes negation and resistance. There is no simple and limpid continuity from the one to the other, but rather transformation of the one into the other out of their mutual conflict. We must suppose that mental formation draws its being or identity from the disappearance of the neuronal, born of a sort of blank space that is the highly contradictory meeting point of nature and history. Only an ontological explosion could permit the transition from one order to another, from one organization to another, from one given to another. The neuronal and the mental resist each other and themselves, and it is because of this that they can be linked to one another, precisely because—*contra* Damasio—they do not speak the same language.

One of the great merits of Bergson is to have shown that every vital motion is plastic, which is to say that it proceeds from a simultaneous explosion and creation. Only in making explosives does life give shape to its own freedom, that is, turn away from pure genetic determinism. Take, for example, this passage, from *Spiritual Energy*:

> When we consider the mechanism of voluntary movement in particular, the functioning of the nervous system in general, and in fact life itself in what is essential to it, we are led to the conclusion that the invariable contrivance of consciousness, from its most humble origin in elementary living forms, is to convert physical determinism to its own ends, or rather to elude the law of the conservation of energy while obtaining from matter a fabrication of explosives, ever intenser and more utilizable. It will then require an almost negligible action, such as the slight pressure of the finger on the hair-trigger of a pistol, in order to liberate at the required moment, in the direction chosen, as great an amount as possible of accumulated energy. . . . To make and utilize explosions of this kind seems to be the unvarying and essential preoccupation of life, from its first apparition in protoplasmic masses, deformable at will, to its complete expansion in organisms capable of free actions.[31]

This formative effect of explosions and this formative action of the explosive correspond to the transformation of one motor regime into another, of one device into another, a transformation necessitating a rupture, the violence of a gap that interrupts all continuity.[32] Such are the law and the adventure of energy. It is thus that one must think the transition from the neuronal to the mental, on the model of the transition from the action of storing glycogen in the muscles to the voluntary action effected through these muscles. Energetic explosion is the idea of nature. In passing from one motor to the other, from one energetic device to the other, force simultaneously loses itself and forms itself differently, just as the metamorphic crisis frees a butterfly from its chrysalis. The sculpture of the self is born from the deflagration of an original biological matrix, which does not mean that this matrix is disowned or forgotten but that it cancels itself.

Despite the explosive resonance of the meanings of plasticity, this vision of things obviously does not correspond to a terrorist conception of the constitution of identity. The explosions in question are clearly understood as energetic discharges, creative bursts that progressively transform nature into freedom. To insist on explosive surges is to say that we are not flexible in the sense that all change of identity is a critical test, which leaves some traces, effaces others, resists its own test, and tolerates no polymorphism. Paradoxically, if we were flexible, in other words, if we didn't explode at each transition, if we didn't destroy ourselves a bit, we could not live. Identity resists its own occurrence to the very extent that it forms it.

In the central nervous system, as we have seen, the formative contradiction—formation/explosion—proceeds from a more original contradiction: that between the maintenance of the system, or "homeostasis," and the ability to change the system, or "self-generation." The nervous system, like any system, is self-regulated, self-organized, which means that it expends considerable energy in assuring its maintenance. Basically, in order to preserve itself from destruction, it must keep itself in the same state. Thus it continuously generates and specifies its own organization. "*Homeostasis*," Damasio explains, "refers to the coordinated and largely automated physiological reactions required to maintain steady internal states in a living organism."[33] But every event coming from outside necessarily comes to affect homeostasis and calls upon "another level of cerebral structure," charged with transforming maintenance into a creative ability. In this way, as we have seen, "a face, a melody, a toothache, the memory of an event," demand a first transformation, or account, within the "neural maps," which in turn must be transformed into images or "mental maps." As Jeannerod says:

> the biological function of intentional action ought . . . to be investigated, not as maintaining a constancy, but rather as generating new properties. . . . [This research results in] a reversal of the concept of the relation between organism and environment: a self-regulated structure can only submit to the influence of the environment, while only a structure capable of self-generated activity could impose its own organization. Intentional movement thus becomes the means by which the organism and the environment reciprocally interact, and by means of which the subject constructs its own representation of the real.[34]

But this transition from "homeostasis" to "self-generation" is not made without rupture or gap.

The plasticity that situates subjectivity between maintenance and construction or production of newness is not smooth. The "chain" that leads from elementary life to the autonomy of a free self, capable not only of integrating the disturbances arriving from the exterior without dissolving itself but also of creating itself out of them, of making its own history, is a movement full of turbulence. Homeostatic energy and self-generating energy are obviously not of the same kind. From this perspective, if the brain is really "always caught up in the act of representing to itself its own change," one might suppose, at the very core of the undeniable complicity that ties the cerebral to the psychical and the mental, a series of leaps or gaps.

Reasoned Resilience

The concept of "resilience," taken up and reworked by Boris Cyrulnik, confirms this proposition. Resilience is indeed a logic of self-formation starting from the annihilation of form.[35] It appears as a psychical process of construction, or rather of reconstruction and self-reconfiguration, developed simultaneously against and with the threat of destruction. In studying the cases of certain "problem children"—children held back, mistreated, sick—Cyrulnik reports that some of them developed processes of resilience, possibilities for a *becoming* on the basis of the *effacement of every future*, for a transformation of the trace or mark, and for a historical transdifferentiation. It is as if, in order to return to themselves after the destructive trials they had suffered, these children had to create their own constancy, to self-generate their homeostasis.

But of course these counter-generations themselves necessarily occur through neuronal reconfigurations and, in consequence, through a becoming-mental of these reconfigurations. Far from obeying a simply continuous movement, these reconfigurations and this becoming are made up of ruptures and resistance. The two energies ceaselessly collide within a resilient person. If these individuals were simply "flexible"—that is to say, if the two energies did not collide with one another—they would be not resilient but conciliatory, that is to say, passive. But these individuals are, on the contrary, capable of *changing difference*. Writing of Romanian orphans who made it out of the traumatic hell of the infamous institutions of the Ceaucescu era, Cyrulnik declares, "the traces left in the brain by early lack of affect . . . and social representations . . . confined the Romanian orphans to lower social levels. But orphans whose brain scans showed an inflation of the ventricles and the cortices when they were placed with host families tell us: 'cerebral traces are reparable.' "[36] Traces can change their meaning.

These extreme examples concern us all. We are right to assert that the formation of each identity is a kind of resilience, in other words, a kind of contradictory construction, a synthesis of memory and forgetting, of constitution and effacement of forms. In excluding all negativity from their discourse, in chasing away every conflictual consideration on the transition from the neuronal to the mental, certain neuroscientists cannot, most of the time, escape the confines of a well-meaning conception of successful personality, "harmonious and mature." But we have no use for harmony and maturity if they only serve to make us "scrappers" or "prodigal elders." *Creating resistance to neuronal ideology is what our brain wants, and what we want for it.*

Conclusion: Toward a Biological Alter-globalism

The problem of a dialectic of identity—between fashioning and destruction—poses itself all the more pointedly as global capitalism, currently the only known type of globalization, offers us the untenable spectacle of a simultaneity of terrorism (daily detonations—in Israel, Iraq, Indonesia, Pakistan . . .) and of fixity and rigidity (for example, American hegemony and its violent rigorism). It is as though we had before our eyes a sort of caricature of the philosophical problem of self-constitution, between dissolution and impression of form. Fashioning an identity in such a world has no meaning except as constructing of countermodel to this caricature, as opposed simply to replicating it. Not to replicate the caricature of the world: this is what we should do with our brain. To refuse to be flexible individuals who combine a permanent control of the self with a capacity to self-modify at the whim of fluxes, transfers, and exchanges, for fear of explosion.

To cancel the fluxes, to lower our self-controlling guard, to accept exploding from time to time: this is what we should do with our brain. It is time to remember that some explosions are not in fact terrorist—explosions of rage, for example. Perhaps we ought to relearn how to enrage ourselves, to explode against a certain culture of docility, of amenity, of the effacement of all conflict even as we live in a state of permanent war. It is not because the struggle has changed form, it is not because it is no longer really possible to fight a boss, owner, or father that there is no struggle to wage against exploitation. To ask "What should we do with our brain?" is above all to visualize the possibility of saying no to an afflicting economic, political, and mediatic culture that celebrates only the triumph of flexibility, blessing obedient individuals who have no greater merit than that of knowing how to bow their heads with a smile.

One can legitimately suppose, with Damasio, that a poetic activity is at work within the brain. But the brain doesn't tell (itself) just any story. There is a cerebral conflictuality, there is a tension between the neuronal and the mental, there is always the possibility that one or another trace will not convert into an image, that this or that opening will not be made, that this or that neuronal arrangement will not rise to the level of consciousness. The story is complex. We must consider that in a certain sense the brain does not obey itself, that it manufactures events, that there can be an excess in the system, an explosive part that, without being pathological, refuses to obey. We have seen that plasticity allows us to combine the thought of a sculpture of the self with that of transdifferentiation. To exist is to be able to change difference while respecting the difference of change: the difference between continuous change, without limits, without adventure, without negativity, and a formative change that tells an effective story and proceeds by ruptures, conflicts, dilemmas. I did not choose at random the example of stem cells above. What is fascinating about stem cells is that they bring together the origin, as their name indicates, and the future, the capacity for self re-form. Is this not the best possible definition of plasticity: the relation that an individual entertains with what, on the one hand, attaches him originally to himself, to his proper form, and with what, on the other hand, allows him to launch himself into the void of all identity, to abandon all rigid and fixed determination?

We have examined the question of the convertibility of neuronal patterns into mental images and, in consequence, of the genesis of the self starting from the "proto-self." We have shown that this genesis supposes that one could account simultaneously for the transition from one level of organization to another, for the transition from one regime of organization to another, and, finally, for the transition from one organizational given to another. In a word, that one could understand and explain the transformation of a pure biological given into a cultural and historical thing: a free psychical consciousness or identity. We have shown that, by proposing no theory or interpretation of this transformation or this transition—which cannot simply be the result of observation or of objective description—neuroscientific discourse in general exposes itself to ideological risk and offers nothing new to mankind, while plasticity, far from producing a mirror image of the world, is the form of another possible world. To produce a consciousness of the brain thus demands that we defend a biological alter-globalism.

This biological alter-globalism is clearly dialectical, as I have said. It demands that we renew the dialogue, in one way or another, with thinkers like Hegel, who is the first philosopher to have made the word *plasticity* into a concept, and who developed a theory of the relations between nature and mind that is conflictual and contradictory in its essence. Rereading his *Philosophy of Nature* could teach us much about the transition from the biological to the spiritual, about the way the mind is really already a "self [*Selbst*]," a "spirit-nature" at whose core "differences are one and all physical and psychical."[1]

Of course, although Hegel could not yet express himself in the idiom of the "neuronal" and the "mental," his constant preoccupation was the transformation of the mind's natural existence (the brain, which he still calls the "natural soul") into its historical and speculative being. But this transformation is the dialectic itself. If there can be a transition from nature to thought, this is because the nature of thought contradicts itself. Thus the transition from a purely biological entity to a mental entity takes place in the struggle of the one against the other, producing the truth of their relation. Thought is therefore nothing but nature, but a negated nature, marked by its own difference from itself. The world is not the calm prolonging of the biological. The mental is not the wise appendix of the neuronal. And the brain is not the natural ideal of globalized economic, political, and social organization; it is the locus of an organic tension that is the basis of our history and our critical activity.

The elaboration of dialectical thinking about the brain also allows us to escape the strict alternative between reductionism and antireductionism, the theoretical trap within which philosophy too often confines itself. On one side —that of the cognitive sciences, in particular—we find massive affirmation of the possibility of an absolute naturalization of cognition and mental processes. On the other, we find the affirmation of the perfectly transcendental character of thought, irreducible to biological determinations. The dialogue between Changeux and Ricoeur in *What*

Makes Us Think? is a good example of this pair of alterna-tives. According to Ricoeur, neither the knowledge we ac-cumulate about brain functioning nor even our certainty that our mental states are conditioned by neuronal organi-zation teaches us the slightest thing about either ourselves or the way we think.[2] Such a position is clearly untenable. It is not pertinent to think of our neuronal apparatus as a simple physiological substrate of thought. Conversely, nei-ther is it defensible to advocate an absolute transparency of the neuronal in the mental, an easy back-and-forth from the one to the other. A reasonable materialism, in my view, would posit that the natural contradicts itself and that thought is the fruit of this contradiction. One pertinent way of envisaging the "mind-body problem" consists in taking into account the dialectical tension that at once binds and opposes naturalness and intentionality, and in taking an interest in them as inhabiting the living core of a complex reality. Plasticity, rethought philosophically, could be the name of this *entre-deux.*

By sketching an ideological critique of the fundamental concepts of the neurosciences, I have tried to steer the de-bate toward a terrain different from that of the tired alterna-tive between reductionism and antireductionism. As it happens, this also involves an ideological critique of plastic-ity. Indeed, so long as we do not grasp the political, eco-nomic, social, and cultural implications of the knowledge of cerebral plasticity available today, we cannot do anything with it.

Between the upsurge and the explosion of form, subjec-tivity issues the plastic challenge. I have tried to position us at the heart of this challenge, while inviting readers to do what they undoubtedly have never done: construct and en-tertain a relation with their brain as the image of a world to come.

Notes

Introduction: Plasticity and Flexibility—For a Consciousness of the Brain

1. The term *neurosciences* has been used since the 1970s. It covers neurobiology, neurophysiology, neurochemistry, neuropathology, neuropsychiatry, neuroendocrinology, etc.

2. Jean-Pierre Changeux, *Neuronal Man: The Biology of the Mind*, trans. Laurence Garey (New York: Pantheon, 1985), xiii.

3. Ibid., xiv.

4. "Cognitive science forms a vast continent of research that touches on many disciplines: cognitive psychology, artificial intelligence, the neurosciences, linguistics, and philosophy of mind. One even talks today of 'cognitive anthropology' and 'cognitive sociology.' . . . The domains covered (perception, memory, learning, consciousness, reasoning, etc.) are studied on many levels: from their biological bases (cell physiology, brain anatomy, . . .) all the way to the study of 'internal mental states' (representations, mental images, problem-solving strategy)" (*Le cerveau et la pensée: La révolution des sciences cognitives*, ed. Jean-François Dortier [Paris: Sciences Humaines Editions, 1999], 4).

5. MRI stands for "magnetic resonance imaging," and PET for "positron emission tomography." On this topic, see: *Annales d'histoire et de philosophie du vivant* 3, "Le cerveau et les images" (Paris: Institut d'édition Sanofi-Synthélabo, 2000).

6. MAOI stands for "Monoamine oxidase inhibitor," and SSRI for "selective serotonin reuptake inhibitor": Prozac, Paxil, Luvox, Celexa, etc.

7. Boris Cyrulnik uses this notion prominently in his work (see the last chapter of this book).

8. Some examples taken from the hundreds of pages devoted to plasticity on the Internet confirm this: "Plasticité nerveuse" (www.chu.rouen.fr); Institut Pasteur, Cours de développement et plasticité du système nerveux (www.pasteur.fr); Equipe CNRS, "Intégration et plasticité synaptique dans le cortex visuel" (unic.cnrs-gif.fr); "Atelier sur la plasticité cérébrale et modélisation mathématique" (crm-montreal.ca); "Développement et plasticité du système nerveux" (sign7.jussieu.fr); "Développement et plasticité du SNC," licence de sciences cognitives, Université Aix-Marseille (sciences-cognitives.org); "Plasticité et régulation de la neurogenèse dans le cerveau" (Incf.cnrs-mrs.fr); "Groupe plasticité post-lésionnelle," Faculté des sciences et des techniques Saint-Jérôme, Marseille (irme.org).

9. This is particularly so in the magazine *La Recherche*.

10. Changeux, *Neuronal Man*, 247.

11. *Cybernetics* comes from the Greek *kubernan*, to govern. Cybernetics is the science constituted by the group of theories about control, regulation, and communication in living things and machines.

12. See the entry "Plasticity in the Nervous System," in *The Oxford Companion to the Mind*, ed. Richard L. Gregory (Oxford: Oxford University Press, 1987), 623.

13. [Malabou here refers to a set of related words not available in English, which I have therefore left in French in the main text. As we use in English the French form *plastique* to signify plastic explosive material, the French use the English form *plastic* (which otherwise does not occur in French). French also has (at least)

two associated terms: the noun *plastiquage*, meaning the act or event of blowing something up using plastic explosives, and the corresponding verb *plastiquer*.—Trans.]

14. This description is a simplified summary of the remarkably precise description given by Marc Jeannerod in his *Le cerveau intime* (Paris: Odile Jacob, 2002), 47. The axon, which is much longer than the dendrites, is in a certain sense the telegraphic line that transmits messages from one neuron to another, or to the muscle or gland that it serves. The axon and the membrane that surrounds it form the nervous fiber. Each neuron produces electrical signals that are propagated along the axon. The transmission of the signals of one neuron to another across the synapse is generally realized by a chemical substance, the neuromediator.

15. Jeannerod, *Le cerveau intime*, 63.

16. Ibid, 66.

17. Changeux, *Neuronal Man*, 247.

18. *Proto-self* and *neuronal self* are terms used by the neurologist Antonio Damasio; we will return to these terms in the last chapter.

19. Luc Boltanski and Eve Chiapello, *The New Spirit of Capitalism*, trans. Gregory Elliot (London: Verso, 2005), 149.

20. Daniel Dennett, *Consciousness Explained* (Boston: Little, Brown, 1991).

21. [Although English and French both have the word *to fold* (*plier*) in transitive and intransitive senses, English lacks the neat pair *prendre le pli / donner le pli*. The French *prendre le pli*, here literally translated as "take the fold," also appears in the phrase *prendre un mauvais pli*, meaning to develop a bad habit. *Donner le pli*, here literally translated as "give the fold," also means to put a crease in something.—Trans.]

22. In the strong sense of the word *genius*: invention, form giving.

23. A gastropod mollusk also called a "sea slug."

24. [The reference is to Malabou's *The Future of Hegel: Plasticity, Temporality, Dialectic*, trans. Lisbeth During (New York: Routledge, 2004), in which she develops the concept of plasticity found in Hegel's *Phenomenology of Spirit*.—Trans.]

1. Plasticity's Fields of Action

1. This is why the group of synthetic materials that can be molded or modeled (bakelite, cellulose, nylon, polyamide, polyester, resin, silicone, etc.) and cannot regain their initial state after being fabricated are called "plastics." Many of them are rigid following formation and cooling.

2. See Ali Turhan, "Des cellules souches adultes greffées sont reprogrammables," *La Recherche* 365 (June 2003): 18, entry "Plasticité."

3. Brain stem cells, for example, necessarily differentiate themselves into one or another type of cell present in the brain: neurons or glial cells. There is a certain amount of room for maneuvering in differentiation, which is exactly what is meant by multipotence, but it remains limited. Multipotent stem cells produce only a restricted number of cellular types.

4. "Multipotence" and "pluripotence" characterize adult stem cells, which are to be distinguished from embryonic stem cells. The latter are called "totipotent" to the extent that they can develop into practically the entire set of two hundred known types of cells that form a large range of tissues and organs, such as the heart, the pancreas, and the nervous system. Embryonic stem cells are therefore capable of giving birth to a complete individual. There would be much to say on the topic of stem cells, as much about their functioning and the astounding possibilities of autologous grafts they seem to promise (grafts of the organs of patients themselves, regeneration of the individual by himself, in a sense) as about the philosophical analysis of the concepts of difference, reparation, transformation, remodeling of the trace or of the path. But that would be another debate. I bring in stem cells here for only two reasons: to furnish a paradigm of the "open" meaning of plasticity and to allow us to envisage the role of (adult) secondary neurogenesis in the modulation of synaptic efficacy.

5. One speaks of the "navigation" of cells.

6. Changeux, *Neuronal Man*, 126.

7. Ibid., 198–99.

8. Jeannerod, *Le cerveau intime*, 17.

9. Ibid.

10. Changeux, *Neuronal Man*, 217.

11. Jean-Claude Ameisen, *La sculpture du vivant: Le suicide cellulaire ou la mort créatrice* (Paris: Seuil, 1999), 30.

12. Jeannerod, *Le cerveau intime*, 20.

13. Ibid., 21.

14. Ibid., 25–27.

15. Donald Holding Hebb (1904–86) is the author of *The Organization of Behavior: A Neuropsychological Theory* (London: Wiley and Sons, 1949). The term *plasticity* was first used by the great Polish neurologist Jerzy Konorski, who proposed a vision of synaptic functioning quite close to that of Hebb (see his *Conditioned Reflexes and Neuron Organization* [Cambridge: Cambridge University Press, 1948], and *Integrative Activity of the Brain* [Chicago: University of Chicago Press, 1967]).

16. See "La Mémoire," a special edition of *La Recherche* 267 (July-August 1994), especially the articles by Masao Ito ("La plasticité des synapses," 778–85) and by Yves Frégnac ("Les mille et unes vies de la synapse de Hebb," 788–90).

17. Neurotransmitters (acetylcholine, adrenaline) allow the transfer of the nervous signal from one side of the synaptic gap to the other. Chemistry thus takes over from electricity (the order of transmission of the nervous signal is electrical-chemical-electrical).

18. The fifth temporal circumvolution of the brain, which plays a very basic role in the process of memorization. The key to plasticity of the brain and of behavior is the ability to learn and to remember; the hippocampus is a region particularly concerned with these operations. Damage to the hippocampus has particularly serious, and often definitive, effects on cognition and memory.

19. On this topic, see Sue D. Healy, "Plasticité du cerveau et du comportement," in *Plasticité*, ed. Catherine Malabou (Paris: Leo Scheer, 2000), 98–113.

20. On the astounding capacity of connections and dendritic arborizations to change form, see the heading "Plasticité" in "Des neurones pleins d'épines," *La Recherche* 368 (October 2003): 16.

21. If the synapses, particularly those of the hippocampus, did nothing but reinforce themselves under the effects of LTP, they would all very quickly attain a maximal degree of efficacy, and it would then be impossible to encode any new information.

22. Jeannerod, *Le cerveau intime*, 10.

23. Heather Cameron is an investigator in the molecular biology laboratory of the National Institute for Neurological Disorders and Strokes (NINDS/NIH) in Bethesda, Maryland. I cite her article "Naissance des neurones et mort d'un dogme," trans. Phillipe Brenier, *La Recherche* 329 (March 2000): 35. [Though written originally in English, and hence translated into French for *La Recherche*, this article has not been published in English.—Trans.]

24. Ibid., 30. For more on the study in question, see Heather Cameron and Ronald McKay, "Discussion Point: Stem Cells and Neurogenesis in the Adult Brain," *Current Opinion in Neurobiology* 8 (1998): 677–80.

25. Alain Prochaintz, *How the Brain Evolved*, trans. W. J. Gladstone (in collaboration with The Language Service, Inc., Poughkeepsie, N.Y.; New York: McGraw-Hill, 1992).

26. Pierre-Marie Lledo, Patricia Gaspar, Alain Trembleau, "La curieuse partition des nouveaux neurones," *La Recherche* 367 (September 2003): 54–60.

27. Ibid., 60. See also G. Miller, "Singing in the Brain," *Science* 299 (January 31, 2003): 646, on the topic of the neuronal renewal required for bird songs.

28. Jeannerod, *Le cerveau intime*, 69.

29. See "Alzheimer, cerveau sans mémoire," *La Recherche*, n.s. 10 (January-March 2003): 27ff.

30. Pascal Giraux and Angela Sirigu, "Les mains dans la tête," *La Recherche* 366 (July-August 2003): 63.

2. The Central Power in Crisis

1. Henri Bergson, *Matter and Memory*, trans. Nancy Margaret Paul and W. Scott Palmer (New York: Zone, 1988), 30.

2. Marc Jeannerod, *La nature de l'esprit* (Paris: Odile Jacob, 2002), 85–86.

3. On the Bergsonian conception, see also Jean-Pierre Changeux, *Neuronal Man: The Biology of the Mind*, trans. Laurence Garey (New York: Pantheon, 1985), 127.

4. Jeannerod, *La nature de l'esprit*, 18.

5. Artificial Intelligence (AI) is an important field of informatics devoted to constructing "intelligent" programs, which is to say, programs capable of analyzing an environment, resolving problems, making decisions, learning, and perceiving.

6. Jeannerod, *La nature de l'esprit*, 31.

7. Ibid., 137.

8. Gilles Deleuze, *Cinema 2: The Time-Image*, trans. H. Tomlinson and R. Galeta (Minneapolis: University of Minnesota Press, 1989), 211.

9. Changeux, *Neuronal Man*, 83.

10. [In the original French here, *des petites morts cérébrales*, Deleuze plays on the phrase *petit mort*, meaning "orgasm." An alternative translation of this sentence would be: "our lived relationship with the brain becomes more and more fragile, less and less 'Euclidean,' and undergoes little cerebral orgasms."—Trans.]

11. Deleuze, *Cinema 2*, 211, 212.

12. "Forty years after its creation, the results of AI are mixed, to say the least. More and more specialists [after the analyses of Searle] devote themselves to the project of 'weak AI,' as opposed to the 'strong AI' of the early years. The project of strong AI was to discover and reconstruct the way in which man thought, and then to surpass it. The project of 'weak AI' is more modest. It consists in simulating, through engineering, those human behaviors 'generally regarded as intelligent,' without worrying about whether or not humans perform these behaviors in the same way. It is preferable today to speak of software as an 'aid' to creation or to decision making, rather than of machines that will replace the human" (Jean-François Dortier, "Espoirs et réalités de l'Intelligence Artificielle," in *Le cerveau et la pensée: La révolution des sciences cognitives*, ed. Jean-François Dortier [Paris: Sciences Humaines Editions, 1999], 115).

13. "The level of description and explication we need is *analogous* to (but not identical to) one of the 'software levels' of description for computers: what we need to understand is how

human consciousness can be realized in the operation of a *virtual machine*" (Daniel C. Dennett, *Consciousness Explained* [Boston: Little, Brown, 1991], 210). The concept of a "virtual machine" is borrowed from Alan Turing.

14. Ibid., 211.

15. Ibid., 188.

16. Ibid., 225.

17. Deleuze, *Cinema 2*, 206–7.

18. Ibid., 213.

19. Ibid., 317n20.

20. Luc Boltanski and Eve Chiapello, *The New Spirit of Capitalism*, trans. Gregory Elliot (London: Verso, 2005), 90.

21. Ibid., 75.

22. Ibid., 73.

23. Ibid., 149.

24. Ibid., 104.

25. Changeux, *Neuronal Man*, 192.

26. Ibid., 137.

27. Marc Jeannerod, *Le cerveau intime* (Paris: Odile Jacob, 2002), 33.

28. Boltanski and Chiapello, *The New Spirit of Capitalism*, 115.

29. Ibid.

30. Ibid.

31. It is worth noting that in the 1990s the [French] word *cadre* ["team"] was replaced by *manager* [borrowed from English—Trans.].

32. Boltanski and Chiapello, *The New Spirit of Capitalism*, 78–79.

33. "The progress of neurobiology revolutionizes our thinking in affirming that neurons are not specialized," claims Jean-Yves Nau in an article entitled "Les neurosciences découvrent les sources du plaisir sensorielle," *Le Monde*, 31 December 2003, p. 17.

34. Cf. Jeannerod, *La nature de l'esprit*, 94–95.

35. Changeux, *Neuronal Man*, 140–41.

36. Boltanski and Chiapello, *The New Spirit of Capitalism*, 92.

37. Ibid., 145.

38. See ibid., 365ff.: "Testing the exploitation of the immobile by the mobile."

39. Ibid., 112.

40. Ibid., 461.

41. Jean-François Allilaire, professor of psychiatry at the University of Paris VI—Pitié-Salpêtrière Hospital, keynote address to the conference "Dépression et neuroplasticité: Évolution ou revolution?" PSY-SNC Colloquium, Cité des Sciences et de l'Industrie, Paris, November 5–8, 2003.

42. Alain Ehrenberg, *La fatigue d'être soi: Dépression et société* (Paris: Odile Jacob, 1998), 217. The contrary behavior, or "hyperimpulsiveness," also corresponds, despite appearances, to the same phenomenon of disconnection.

43. Ibid., 221.

44. Allilaire, keynote address.

45. Ehrenberg, *La fatigue d'être soi*, 235.

46. Ibid., 234.

47. Ibid., 236.

48. Robert Castel, *From Manual Workers to Wage Laborers: Transformation of the Social Question*, trans. and ed. Richard Boyd (New Brunswick: Transaction Publishers, 2003), xv.

49. Ibid., 3.

50. Boltanski and Chiapello, *The New Spirit of Capitalism*, 363–64.

51. Ibid., 450.

52. Peter Kramer, *Listening to Prozac* (New York: Viking, 1997), 297.

53. We can also think of sclerotic plaque, the adjective *sclerotic* ("fixed, no longer evolving") being the exact contrary to the adjective *plastic*.

3. "You Are Your Synapses"

NOTE: The phrase that forms my chapter title comes from the American neurologist Joseph LeDoux, in his book *Synaptic Self*

(New York: Viking, 2002): "The central message of this book is 'You are your synapses'" (ix). This book is currently one of the most complete, most clear, and most interesting on the topic of cerebral functioning.

1. Antonio R. Damasio, *The Feeling of What Happens: Body and Emotions in the Making of Consciousness* (New York: Harcourt, Brace & Co., 1999), 313.

2. LeDoux, *Synaptic Self*, 13.

3. Ibid., 174.

4. Ibid., 2.

5. Ibid., 173.

6. Ibid., 304.

7. Damasio, *The Feeling of What Happens*, 22.

8. Ibid., 170.

9. Ibid., 22. [Damasio's French translators have rendered "core self" as *soi central*.—Trans.]

10. Ibid., 174 [Table 6.1.]

11. Ibid., 154.

12. The psychoanalyst André Green has further remarked, in his book *La causalité psychique: Entre nature et culture* (Paris: Odile Jacob, 1995), that "the presence of the concept of representation is almost synonymous with that of psychism" (314).

13. In the course of fascinating clinical analyses, Damasio shows that "the brain knows more than the conscious mind reveals," as is witnessed by certain illnesses attending serious memory lesions, in which the "proto-self" remains intact. See Damasio, *The Feeling of What Happens*, 42ff. One can therefore conclude that "the power to make neural patterns . . . is preserved even when consciousness is no longer being made" (166).

14. Ibid., 230. [Damasio's French translators have rendered "conscience" as *conscience morale*.—Trans.]

15. Ibid., 170.

16. Ibid., 172.

17. Ibid., 9.

18. LeDoux, *Synaptic Self*, 3.

19. Damasio, *The Feeling of What Happens*, 169.

20. Ibid., 228.

21. Jean-Pierre Changeux and Alain Connes, *Conversations on Mind, Matter, and Mathematics*, trans. M. B. DeBevoise (Princeton, N.J.: Princeton University Press, 1998), 113.

22. Damasio, *The Feeling of What Happens*, 225.

23. Ibid., 223.

24. For a critique of mental Darwinism, see Green, *La causalité psychique*, 26ff.

25. Gilles Deleuze, *Cinema 2: The Time-Image*, trans. H. Tomlinson and R. Galeta (Minneapolis: University of Minnesota Press, 1989), 204–5.

26. Examples cited by Jean Decety in his article "Les images du cerveau: Intérêt et limites des techniques de neuro-imagerie," *Annales d'histoire et de philosophie du vivant*, vol. 3, *Le cerveau et les images* (Paris: Institut d'édition Sanofi-Synthélabo, 2000), 39.

27. Alain Ehrenberg, *La fatigue d'être soi: Dépression et société* (Paris: Odile Jacob, 1998), 261.

28. Cf. the proposal of Joseph LeDoux in *Synaptic Self*: "it can also improve quality of life, as when it uncovers new ways of treating neurological or psychiatric disorders" (3). Or again: "A key question you may be asking yourself is whether all this hardcore neuroscience has, in fact, any practical application. In other words, might it be possible to use this kind of work to help improve normal memory and, ever more important, to rescue or prevent age-related memory loss?" (172).

29. The proto-self is plastic to the extent that, as Damasio says, it "does not occur in one place only, and it emerges dynamically and continuously out of multifarious interacting signals that span varied orders of the nervous system" (*The Feeling of What Happens*, 154).

30. Cf. Sigmund Freud, "Project for a Scientific Psychology" (1950 [1895]), in *The Standard Edition of the Complete Psychological Works of Sigmund Freud*, vol. 1, *1886–1899: Pre-Psycho-Analytic Publications and Unpublished Drafts*, trans. James Strachey (London: Hogarth Press, 1966), 281–391.

31. Henri Bergson, *Spiritual Energy*, trans. H. Wildon Carr (New York: Henry Holt, 1920), 44–45.

32. These energetic ruptures, as I have noted above, can also be seen at the most rudimentary level of the constitution of neuronal configurations, which requires the conversion of electrical signals into chemical signals and then back again to electrical signals. See LeDoux, *Synaptic Self*: "The full sequence of communication between neurons is thus usually electrical-chemical-electrical: *electrical* signals coming down axons get converted into *chemical* messages that help trigger *electrical* signals in the next cell. . . . As hard as it may be to imagine, electrochemical conversations between neurons make possible all of the wondrous (and sometimes dreadful) accomplishments of human minds. Your very understanding that the brain works this way is itself an electrochemical event" (47–48).

33. Damasio, *The Feeling of What Happens*, 39.

34. Marc Jeannerod, *La nature de l'esprit* (Paris: Odile Jacob, 2002), 111. On the topic of self-generation, see also Changeux, *Neuronal Man*: "Moreover, the human brain can develop strategies on its own. It anticipates coming events and elaborates its own programs. This capacity for self-organization is one of the most remarkable features of the human cerebral machine, and its supreme product is thought" (127).

35. The concept of "resilience" comes from material physics. "Resilient" means: "what resists impacts (more or less), what is characterized by a greater or lesser resilience." "Resilience": "relation between kinetic energy absorbed to cause a break in a metal to the surface of a broken section. Resilience (in kg per cm²) characterizes resistance to shock or impact."

36. Boris Cyrulnik, *Un merveilleux malheur* (Paris: Odile Jacob, 1999).

Conclusion: Toward a Biological Alter-globalism

1. G. W. F. Hegel, *Philosophy of Mind*, trans. W. Wallace and A. V. Miller, rev. and introd. Michael Inwood (Oxford: Oxford University Press, 2007), § 396, 53.

2. Jean-Pierre Changeux and Paul Ricoeur, *What Makes Us Think? A Philosopher and a Scientist Argue about Ethics, Human Nature, and the Brain*, trans. M. B. DeBevoise (Princeton, N.J.: Princeton University Press, 2002).

Perspectives in
Continental Philosophy Series
John D. Caputo, series editor

Jean-Luc Marion, *In Excess: Studies of Saturated Phenomena*.- Translated by Robyn Horner and Vincent Berraud.

Phillip Goodchild, *Rethinking Philosophy of Religion: Approaches from Continental Philosophy*.

William J. Richardson, S.J., *Heidegger: Through Phenomenology to Thought*.

Jeffrey Andrew Barash, *Martin Heidegger and the Problem of Historical Meaning*.

Jean-Louis Chrétien, *Hand to Hand: Listening to the Work of Art*. Translated by Stephen E. Lewis.

Jean-Louis Chrétien, *The Call and the Response*. Translated with an introduction by Anne Davenport.

D. C. Schindler, *Han Urs von Balthasar and the Dramatic Structure of Truth: A Philosophical Investigation*.

Julian Wolfreys, ed., *Thinking Difference: Critics in Conversation*.

Allen Scult, *Being Jewish/Reading Heidegger: An Ontological Encounter*.

Richard Kearney, *Debates in Continental Philosophy: Conversations with Contemporary Thinkers*.

Jennifer Anna Gosetti-Ferencei, *Heidegger, Hölderlin, and the Subject of Poetic Language: Towards a New Poetics of Dasein*.

Jolita Pons, *Stealing a Gift: Kirkegaard's Pseudonyms and the Bible*.

Jean-Yves Lacoste, *Experience and the Absolute: Disputed Questions on the Humanity of Man*. Translated by Mark Raftery-Skehan.

Charles P. Bigger, *Between* Chora *and the Good: Metaphor's Metaphysical Neighborhood*.

Dominique Janicaud, *Phenomenology "Wide Open": After the French Debate*. Translated by Charles N. Cabral.

Ian Leask and Eoin Cassidy, eds., *Givenness and God: Questions of Jean-Luc Marion*.

Jacques Derrida, *Sovereignties in Question: The Poetics of Paul Celan*. Edited by Thomas Dutoit and Outi Pasanen.

William Desmond, *Is There a Sabbath for Thought? Between Religion and Philosophy*.

Bruce Ellis Benson and Norman Wirzba, eds. *The Phenomoenology of Prayer*.

S. Clark Buckner and Matthew Statler, eds. *Styles of Piety: Practicing Philosophy after the Death of God*.

Kevin Hart and Barbara Wall, eds. *The Experience of God: A Postmodern Response*.

John Panteleimon Manoussakis, *After God: Richard Kearney and the Religious Turn in Continental Philosophy*.

John Martis, *Philippe Lacoue-Labarthe: Representation and the Loss of the Subject*.

Jean-Luc Nancy, *The Ground of the Image*.

Edith Wyschogrod, *Crossover Queries: Dwelling with Negatives, Embodying Philosophy's Others*.

Gerald Bruns, *On the Anarchy of Poetry and Philosophy: A Guide for the Unruly*.

Brian Treanor, *Aspects of Alterity: Levinas, Marcel, and the Contemporary Debate*.

Simon Morgan Wortham, *Counter-Institutions: Jacques Derrida and the Question of the University*.

Leonard Lawlor, *The Implications of Immanence: Toward a New Concept of Life*.

Clayton Crockett, *Interstices of the Sublime: Theology and Psychoanalytic Theory*.

Bettina Bergo, Joseph Cohen, and Raphael Zagury-Orly, eds., *Judeities: Questions for Jacques Derrida*. Translated by Bettina Bergo, and Michael B. Smith.

Jean-Luc Marion, *On the Ego and on God: Further Cartesian Questions*. Translated by Christina M. Gschwandtner.

Jean-Luc Nancy, *Philosophical Chronicles*. Translated by Franson Manjali.

Jean-Luc Nancy, *Dis-Enclosure: The Deconstruction of Christianity*. Translated by Bettina Bergo, Gabriel Malenfant, and Michael B. Smith.

Andrea Hurst, *Derrida Vis-à-vis Lacan: Interweaving Deconstruction and Psychoanalysis*.

Jean-Luc Nancy, *Noli me tangere: On the Raising of the Body*. Translated by Sarah Clift, Pascale-Anne Brault, and Michael Naas.

Jacques Derrida, *The Animal That Therefore I Am*. Edited by Marie-Louise Mallet, translated by David Wills.

Jean-Luc Marion, *The Visible and the Revealed*. Translated by Christina M. Gschwandtner and others.

Michel Henry, *Material Phenomenology*. Translated by Scott Davidson.

DATE DUE
